# JOURNAL OF SC ‖‖‖‖‖‖‖‖‖‖‖‖‖‖ ION

S0-BYN-382

## A Publication of the [ ] [ ] ration

(ISSN 0892-3310) published quarterly, and continuously since 1987

SUBSCRIPTIONS & PREVIOUS JOURNAL ISSUES: Order forms on back pages or at scientificexploration.org.

**Society for Scientific Exploration Website**—https://www.scientificexploration.org

*Journal of Scientific Exploration* (ISSN 0892-3310) is published quarterly in March, June, September, and December by the Society for Scientific Exploration, 12 Candlewood Drive, Petaluma, CA 94954 USA. Society Members receive online *Journal* subscriptions with their membership. Online Library subscriptions are $165.

# JOURNAL OF SCIENTIFIC EXPLORATION
## A Publication of the Society for Scientific Exploration

**AIMS AND SCOPE:** The *Journal of Scientific Exploration* is an Open Access journal, which publishes material consistent with the Society's mission: to provide a professional forum for critical discussion of topics that are for various reasons ignored or studied inadequately within mainstream science, and to promote improved understanding of social and intellectual factors that limit the scope of scientific inquiry. Topics of interest cover a wide spectrum, ranging from apparent anomalies in well-established disciplines to rogue phenomena that seem to belong to no established discipline, as well as philosophical issues about the connections among disciplines. The *Journal* publishes research articles, review articles, essays, commentaries, guest editorials, historical perspectives, obituaries, book reviews, and letters or commentaries pertaining to previously published material.

https://doi.org/10.31275/2018.1373B for this whole issue, JSE 32:3, Fall 2018, and https://doi.org/10.31275/2018.1373 for the Whole Issue PDF file.

# JOURNAL OF SCIENTIFIC EXPLORATION
## A Publication of the Society for Scientific Exploration

Volume 32, Number 3                                                          2018

## SSE NEWS

*Journal of Scientific Exploration*, Vol. 32, No. 3, pp. 489–494, 2018    0892-3310/18

DOI: https://doi.org/10.31275/2018/1363

In my Editorial in the last issue, I dealt at some length with the topic of experimental replicability, revisiting a subject I'd addressed in another Editorial five years earlier. And back then, I followed that initial Editorial with another, dealing with an important and too often neglected side-issue— namely, whether (or to what extent) we should consider scientific expertise to be an art, or something more like a gift than a skill. As far as I can tell, this interesting topic continues to receive even less attention than the usual concerns over replicability. So now I'd like once again to raise the relevant issues. Perhaps the second time is a charm.

Philosopher Karl Popper notoriously once wrote: "Any empirical scientific statement can be presented (by describing experimental arrangement, etc.) in such a way that *anyone who has learned the relevant techniques* can test it" (Popper 1959/2002:99, emphasis added). In my Editorial last issue, I noted that given the inevitable differences between original experiments and replication attempts—magnified in the behavioral sciences (and parapsychology) by many additional kinds of potentially relevant variables (such as well-documented experimenter effects), it may be unreasonable to expect success when replication attempts are conducted by someone other than the original experimenter. That point is relatively familiar. What I want to consider more closely now are the less familiar, related questions: What are the relevant techniques? Can they be captured and conveyed by a mere list of procedures, like a recipe for baking bread? And in particular: To what extent can these techniques even be learned?

When we consider what makes a good physician, psychiatrist, or clinical psychologist, we recognize that a key requirement is something that no mere recipe can capture adequately and that can't easily be taught (if it can be taught at all)—namely, having a "nose" so to speak for what matters—e.g., diagnostically relevant clues. Granted, education can help point one in the right direction, but it can't turn just anyone into a great diagnostician, or a great detective, any more than it can turn just anyone into a great human being. Indeed, one would think that another key requirement of these professions is the ability to relate successfully to others—that is, to have the kind and degree of sensitivity, empathy, or whatever exactly is needed, to understand what others are saying (e.g., to know what's *behind* their words), to know when they're dissembling or withholding information, to make them feel comfortable, supported, etc. And that, too, is something that's very difficult to teach, if it can be taught at all. Very likely, it requires native aptitudes that people simply either have or lack—the qualities in virtue of which some are especially good in relating to other people. To think that these qualities can be acquired merely through education is as foolish as thinking that through formal education alone one can learn to be compassionate, courageous, or witty—or more generally, that one can change deep features of one's character. Similarly, it would be astonishing (if not miraculous) if scientific expertise generally and experimental expertise specifically (perhaps especially in the biological and behavioral sciences) didn't likewise require certain aptitudes or native capacities with which only some are fortunately endowed. And that may also include having a nose for what matters.

Although this bit of commonsense wisdom may frequently be overlooked, it's hardly a new observation. Perhaps the origins trace back as far as Plato's *Republic*. Plato was concerned with (among other things) what human excellence amounted to, and he noted that this must be answered relative to the different roles that a person can fulfill—for example, that of a teacher, parent, musician, military commander, boxer. A person isn't simply excellent *simpliciter*. That's why we can say that someone (for example) is a good musician but a lousy parent. Plato also noted that we can evaluate someone *qua* (i.e. in the capacity of a) person—along some kind of moral dimension. Indeed, we can say that someone is a good person but a terrible teacher (an all too common phenomenon, in fact), or a good computer technician but a rotten human being.

Now Plato had his own philosophical and political agenda in writing *The Republic* and so he didn't extend his observations in the directions that concern me here. But we can note that excellence in a person's various capacities might be related in intimate (perhaps even lawlike) ways to excellence in some other capacities. For example, it's likely that a scientist's personal qualities (e.g., character traits) could be a deciding factor in determining whether experiments succeed or fail, or whether theory-building and data-gathering are productive. And I don't have in mind only such relatively coarse measures as (say) whether a parapsychologist is a sheep (believer) or goat (non-believer or skeptic) (see, e.g., Wiseman & Schlitz 1997). Some examples will illustrate what I have in mind. (I'll confine my comments to work in parapsychology, but I encourage readers to find analogues in other areas of science.)

When I began my serious study of parapsychological research back in the 1970s, I was struck by the following episode at a conference of the

Parapsychological Association. One of the presenters was Helmut Schmidt, an exceptionally creative and successful theoretician and experimenter. Helmut gave a talk in which he described his latest success in testing subjects' ability to influence the output of random number generators. Helmut's talk was given with his usual (and considerable) energy and enthusiasm. For example, he described in a very animated way how he encouraged his subjects to imagine themselves psychically *pushing* the RNG. And the word "pushing" he expressed with great emphasis and dramatic gestures.

Following this presentation was a talk given by a young woman who had tried unsuccessfully to replicate one of Schmidt's earlier experiments. I know from having spoken to her that she was a very pleasant person. But her personality was so different from that of Schmidt, one could be forgiven for thinking that the two experimenters were members of different species. Helmut was charismatic, extroverted, enthusiastic, and dynamic. It was easy to see how he could have effectively encouraged his subjects to succeed. By contrast, this young woman was relatively lifeless, monotonous, and insipid. Her talk was given with an almost total lack of affect and vocal inflection, and that wasn't just a matter of stage-fright; that was her manner of talking. So it was equally easy to see how she might have failed to inspire or excite her subjects. Similarly, perhaps the late John Beloff's notoriously poor track record in conducting or supervising successful psi experiments connects with his gentle and quite understated personality, despite the fact that he clearly qualified as a sheep—that is, despite his demonstrated sympathy for psi research and his obvious conviction about the positive merits of the best cases.

Along the same lines, in both psi research and the behavioral sciences generally, experimental success might require, in addition to (or instead of) charisma, a supportive experimental personality that can make subjects feel safe or comfortable about participating in the experiment, and which can help them trust the experimenter. Many believe (as I do) that this is why Russell Targ (another low-key personality) has been so consistently successful in conducting remote viewing trials. And clearly, only some people have that kind of character trait. Moreover, it may also be a matter of the way personality styles *fit* with one another. Even a generally supportive or encouraging person may still rub some people the wrong way, if their personalities are broadly incompatible. After all, that's one reason we can feel comfortable in life with certain people but not others.

Now you might think that psychologists especially should be keenly aware of these sorts of interactions and potential personality conflicts. I used to think so—at least I did early in my academic career, before I began to meet more and more university psychologists and started attending their parties (clinical psychologists, arguably, are a different kind of animal). At that point, however, I realized that my hosts often had almost no idea which people should be invited together to the same affair, and which people would almost certainly create friction when placed in a common environment. I could only wonder, then, how that ironic blindness might also affect their professional work—for example, their ability to relate to their subjects, or to select appropriate graduate assistants to interact with their subjects.

Not surprisingly, there has been some mainstream research on the personality correlates to successful experimentation in psychology. But those I've seen have been rather (and all too commonly) superficial, focused on such relatively rudimentary measures as, for example, experimenter need for social influence, experimenter desire for control, and subject need for social approval (see, e.g., Hazelrigg, Cooper, & Strathman 1991). They seldom rise above commonsense, very general conjectures and observations that probably never needed to be confirmed with the aid of precious research funds. Moreover, as far as the study cited above is concerned, given the authors' own experimental procedures, one can only wonder how they evaluated the relevance of their own personality traits in leading to their results. That is, one can only wonder about the wisdom of experimentally investigating experimental biasing—at least, in the absence of detailed and reliable information about the experimenters' own character traits. Personally, I suspect that experimentation is simply not the way to proceed here. Probably, there's much more to learn from keen and sensitive observers' careful and penetrating examination of successful experimenters, as well as of subjects who succeed at psi tasks under a wide range of experimental conditions.

I mentioned earlier that scientists might need a "nose" (or perhaps an "eye") for relevant data, and that in the absence of this ability their work might exhibit systematic deficiencies. This is a criticism I've lodged many times against the postmortem survival research of Ian Stevenson. Please don't misunderstand me on this; I believe that Stevenson's work constitutes a monumentally important and valuable repository of data. However, as I've argued in detail (see, e.g., Braude 2003), Stevenson repeatedly treated the subjects of his case investigations as if they were psychological stick figures, with little or no depth or breadth of personality, and as if they had no deeply hidden issues guiding their lives in the subtle ways most of us can discover from our usual life blunders and successes—for example, the cunning and often indirect or elusive ways we might repeatedly entangle ourselves in lethal relationships, or undermine our attempts to succeed professionally.[1] Consider, for example, the blatant clues about motivations and subject psychopathology Stevenson missed in the well-known case of Sharada (Braude 2003:Chapter 4). For all his many virtues, I'd say Ian was blind to much of what really deserved his attention. And as a result, he repeatedly underestimated the power of sophisticated and reasonable alternatives to the hypotheses of reincarnation specifically and survival generally.

Now if it's true that scientific success or failure sometimes hinges on the presence or absence of certain personality traits of the scientist and (in the case of experiments) is not simply a matter of following a recipe of procedures, what can be done about this? It seems unlikely that graduate programs in the sciences will suddenly—or ever—award advanced degrees only to students passing a battery of relevant psychological tests. And it seems equally unlikely that scientists will volunteer themselves for psychological profiling, the results of which can be published alongside their research. For example, members of a parapsychological insiders' listserve to which I subscribe showed very little enthusiasm for the idea, even though listserve members have floated proposals for publishable psychological screening of experimenters several times over the years.

Moreover, I suspect that many (most?) scientists like to perpetuate the myth that they're especially objective observers and agents, and not the steaming, stinky cauldrons of fears, insecurities, flaws, and issues that afflict everyone else. Perhaps the most we can hope for is a rejection of Popper's simplistic statement about scientific expertise, a correspondingly more sophisticated assessment of experimental results, and a willingness to consider seriously the full range of variables (including character traits) that can affect experimental outcome. And more generally, we can perhaps hope for a greater appreciation of the fact that scientists, like other human beings, have both personalities and feelings, and that they're subject to the same grubby concerns, frailties, and life issues that influence even the most mundane actions. Perhaps then we'll see a wider acknowledgment that scientific success and character traits are not neatly separable. And who knows, perhaps then we'll see a more sensible appraisal of replication attempts in areas of frontier science.

—STEPHEN E. BRAUDE

## Note

[1] For an exemplar of a more penetrating way to consider the behavior of both experimenters and subjects, see Eisenbud (1992).

## References Cited

Braude, S. E. (2003). *Immortal Remains: The Evidence for Life after Death*. Lanham, MD: Rowman & Littlefield.

Eisenbud, J. (1992). *Parapsychology and the Unconscious*. Berkeley, CA: North Atlantic Books.

Hazelrigg, P. J., Cooper, H., & Strathman, A. J. (1991). Personality moderators of the experimenter expectancy effect: A reexamination of five hypotheses. *Personality and Social Psychology Bulletin, 17*(5):569–579.

Popper, K. (1959/2002). *The Logic of Scientific Discovery*. New York: Harper.

Wiseman, R., & Schlitz, M. (1997). Experimenter effects and the remote detection of staring. *Journal of Parapsychology, 61*(3):197–207.

*Journal of Scientific Exploration*, Vol. 32, No. 3, pp. 495–513, 2018     0892-3310/18

*RESEARCH ARTICLE*

# The Effects of Geophysical Anomalies on Biology

LYDIA GIANNOULOPOULOU

Department of Physiology, University of Ioannina, Greece
lydarch@tee.gr

ANGELOS EVANGELOU

Faculty of Medicine, University of Ioannina, Greece

SPYROS KARKABOUNAS

Faculty of Medicine, University of Ioannina, Greece

STAVROS PAPAMARINOPOULOS

Department of Geology, University of Patras, Greece

Submitted February 7, 2018; Accepted May 12, 2018; Published June 30, 2018

DOI: https://doi.org/10.31275/2018.1295
Copyright: Creative Commons CC BY NC ND

**Abstract**—The effect of location and its geophysical properties on biology has been known since ancient times. This paper makes an attempt to define geophysical anomalies, analyze the various parameters that constitute them, and analyze the mechanisms through which these anomalies interact with human biology, flora, and fauna.

*Keywords*: geophysical anomalies—earth radiation—magnetic field—
        bioelectromagnetism

## Introduction

The science of geobiology concerns the study of the ways in which a site and its various characteristics affect the development of the biosphere.

Hippocrates, the father of Medicine, in his work *Concerning Nutrition*, mentions that the geographic location and ground typology are equally important health factors as nutrition is. He states (Figure 1):

> The success of proper diagnosis lies in the fact that the physician must know the nature of man as a whole. Human health, in order to be achieved and maintained, needs proper nutrition in proportion to gender, age, work, yearly season, climatic change, in sync with the geographical location of the place where one lives and the prevailing conditions, the typology of the soil, and finally the influences of the sun, moon, and universe in our lives. (Hippocrates 1992: Paragraph 2 [Ιπποκράτης, Περί Διατροφής, παρ. 2])

Δεῖ δε, ὡς ἔοικε, τῶν πόνων διαγινώσκειν τὴν δύναμιν καὶ τῶν κατὰ
φύσιν καὶ τῶν διὰ βίης γινομένων, καὶ τινες αὐτῶν ἐς αὔξησιν
παρασκευάζουσι σάρκας καὶ τινες ἐς ἔλλειψιν, καὶ οὐ μόνον ταῦτα,
ἀλλὰ καὶ τὰς ξυμμετρίας τῶν πόνων πρὸς τὸ πλῆθος τῶν σιτίων καὶ
τὴν φύσιν τοῦ ἀνθρώπου καὶ τὰς ἡλικίας τῶν σωμάτων, καὶ πρὸς τὰς
ὥρας τοῦ ἐνιαυτοῦ καὶ πρὸς τὰς μεταβολὰς τῶν πνευμάτων, καὶ πρὸς
τὰς θέσεις τῶν χωρίων ἐν οἷσι διαιτέονται, πρός τε τὴν κατάστασιν
τοῦ ἐνιαυτοῦ. Ἄστρων τε ἐπιτολὰς καὶ δύσιας γινώσκειν δεῖ, ὅκως
ἐπίστηται τὰς μεταβολὰς καὶ ὑπερβολὰς φυλάσσειν καὶ σίτων καὶ πο
τῶν καὶ πνευμάτων καὶ τοῦ ὅλου κόσμου, ἐξ ὧν περ αἱ νοῦσοι τοῖσιν
ἀνθρώποισι φύονται.

**Figure 1. Original text from *Concerning Nutrition* by Hippocrates in ancient Greek.**


It has been observed that the various features of the subsoil, its morphology, and its composition, may create beneficial or adverse effects on humans, flora, and fauna.

Architecture, as currently practiced, does not take into account the possible effect of the ground from the point of view of geophysical anomalies created by various geological features.

Earth radiation anomalies are related to various phenomena, such as the piezoelectricity, radioactivity, geochemical gases, seismic faults, gravity anomalies, electromagnetic emission, electro-kinetic phenomena of subterranean water flow, spinning electric fields, ion flow, conductivity discontinuities, non-dipolar magnetic fields, geoplasma, and geoneutrons. Specific geological materials can also alter and/or augment these phenomena.

This paper considers the physical aspects of these anomalies, and the way they can interact with biology—in particular how they can affect humans, animals, and plants.

The human body, via different parameters, can interact with these phenomena, due to the existence of magnetite, iron, silica, and water, and also its own electromagnetic fields. Research has shown that long-term exposure at geologically stressed locations can have diverse negative effects on people, especially on brain activity, blood health, and the immune system.

Various similar effects have been noted on flora and fauna. There are studies that have shown the significant effects of these anomalies on laboratory animals with clear results focused on malignancies.

Furthermore, there are indications that different types of geophysical anomalies can affect plants, and various parameters are responsible for the result in yield, weight, shape, and time growth.

## Geophysical Anomalies

### *Classification*

*Geophysical anomalies* are the zones within which significant changes take place in the parameters of various fields, such as the natural magnetic, gravitational, and electric fields. These are locally distorted due to underlying geological features, and the resulting anomalies are measurable. The existence of some local differentiation of the subsoil can be scientifically analyzed and gives an explanation of the results.

The specific phenomena associated with geobiology are the following.

**Piezoelectric or piezomagnetic effect**. Concerns the electrical charge accumulated in certain solid materials (such as crystals with defects, certain ceramics, and biological matter such as bones, DNA, and various proteins) as a reaction when mechanical stress is applied (Hacker, Pauser, & Augner 2011, Freund 2003, Adler, Le Mouel, & Zlotnicki 1999).

**Hydrogeophysical phenomenon** is the low-intensity electric current, and hence the magnetic field associated with it, created by the friction between groundwater and porous limestone rocks, as the water molecules attempt to pass though the rock's micro tunnels within its matrix (Burke & Halberg 2005, Yang, Kostiuk, & Kwok 2003, Adler, Le Mouel, & Zlotnicki 1999).

**Radioactivity** is the presence of radioactive materials such as granite, and others, that create conditions for the emission of radioactivity (United Nations Scientific Committee 1993).

**Geochemical gases** are the release of geochemical gases, such as radon caused by a variety of complex geological conditions within a severe seismogenic environment (United Nations Scientific Committee 1993).

**Ion flow** concerns an upward flow of positive or negative ions emanating from geological formations (Burke & Halberg 2005).

**Strong magnetic, electrical, or electromagnetic anomaly** concerns the strong local difference in the magnetic, electrical, or electromagnetic properties of the underlying geological structures, which differ in magnetic susceptibility and electrical conductivity, respectively (Florinsky 2010, Burke & Halberg 2005, Persinger 1987).

**Strong gravity anomaly** is the sudden local variation of the gravitational field intensity, and is related to the change in density of adjacent geological structures (Florinsky 2010, Gak & Gridin 2008).

**Tectonic faults** are the creation of an electromagnetic field in a pre-seismic period (Florinsky 2010, Shitov 2006, Persinger 1987).

**Non-dipolar magnetic field** concerns the flow of a non-dipolar magnetic field (Miller & Lonetree 2013).

**Ground electric potential anomalies**. Concerns the creation of electrical potential in contacts of geological structures with different electrical properties (Burke & Halberg 2005).

**Geoplasma**. Concerns the emanation of low energy plasma related to geo-anomalies and heterogeneities in the structure of geophysical fields (Mamirova 2010).

**Geoneutrons**. Concerns anomalies in the flow of neutrons emanating from geological structures (Langer 2008, 1997).

### Geo-Electric Phenomena

The Earth's electromagnetic field is a result of the interaction between the magnetic field originating from the melted iron–nickel core of the planet and the charged ionosphere gases. The diurnal fluctuations in the geomagnetic field depend on the solar motion and the solar winds. It also varies according to the lunar day and the month, and annually depending on the distance from the sun.

The Earth's surface and the ionosphere create an electrodynamic cavity producing micro-pulses in the magnetic field in extremely low frequencies, from about 7.83 Hz to 25 Hz (Schumann Resonances). Most of the micro-pulses energy is concentrated at about 10 Hz. Solar flares disperse charged particles in the earth's field, causing magnetic storms. These particles have already been incorporated in the outer boundaries of the field (the Van Allen belt), which protects us from their absorption and other high energy cosmic rays (Becker & Selden 1985).

The phenomenon of induction refers to the change in the intensity of the magnetic field, which will generate an electrical current in any conductive substance present. Consequently, the daily fluctuations in the intensity of the geomagnetic field produce the so-called telluric currents running through the ground near the surface.

These telluric currents produce their own magnetic field, which will amplify or weaken the geomagnetic field according to its polarization. The change in the magnetic field is proportional to the change in the telluric current's intensity, following a ratio known in science (Hessler & Wescott 1959).

### Geological Phenomena

When two different types of geological sub-rock or even sub-soil are in contact, a conduction discontinuity is created that can weaken or potentiate the daily magnetic fluctuations, sometimes more than one hundredfold (Rikitake & Honkura 1986).

This change in the magnetic field creates additional electrical charges, so in these areas the electric currents of the ground are much higher than the surrounding area. Telluric electric currents attract electrically charged air molecules of the opposite charge.

Rocks such as basalt, volcanic lava, limestone, granite, and others with a high content of clay, magnetite, iron, magnesium, or other metals exhibit high electrical conductivity, thus becoming channels for any electric current, such as the one created by an underground water flow. In particular, some types of granite typically emit radon and neutrons. Radon and radiation create ions (Burke & Halberg 2005).

Copper-rich rocks are particularly conductive, as copper is one of the most electrically conductive metals. Quartz-rich rocks create the piezo-electric phenomenon, as they have the ability to store electrical charges, produced in their crystal lattices' defects due to tectonic stresses, like no other mineral; they are therefore used in watch construction. Furthermore, the soil's ability to conduct electricity is proportional to its water content (Burke & Halberg 2005).

The presence of limestone has a strong interaction with water flow, as its geological interstructure is perfect for the production of natural electricity by hydro-geophysical means as described below in the next section. Electrons are removed from the rainwater as it passes through the porosity of the rock and are attached to it, in a process called adsorption (Mizutani et al. 1976).

### Hydro-Geophysical Phenomena

Therefore, water molecules now have a positive charge and have left the chalk with a negative charge. The phenomenon becomes double-strengthened as water dissolves in chalk. Calcium carbonate molecules will be broken down; during this phenomenon, the calcium rock is charged negatively by the free electrons extracted from the water molecule, leaving the water molecules positively charged (Burke & Halberg 2005).

The overall result is that chalk has a significant negative charge and the flowing water is positively charged. As we know, the opposite is attracted and thus an electric charge is created in the ground. The movement of water by itself will create a magnetic field the fluctuation of which depends on the porosity of the rock, and chalk is extremely porous (Martin, Haupt, & Greenfield 1982).

Current densities of $2.5 * 10^{-4}$ amp/m$^2$ can be generated by constant groundwater migration in artesian aquifers. These DC currents are sufficient to generate magnetic anomalies exceeding 200 nT in both the Menindee Trough (Australia) and the Karoo Basin (South Africa). Telluric currents associated with ionospheric activity can be detected by variations in magnetic induction, but constant offsets associated with streaming potentials and geochemical activity (SP anomalies) are obscured by noise. Consequently, some regional magnetic anomalies may be wrongly attributed to variations in magnetic susceptibility with residuals explained by remanence (Cull & Tucker 1986).

Scientists studying the La Fournaise volcano near Madagascar found that underground water flow through volcanic rocks can generate electricity, too (Adler, Le Mouel, & Zlotnicki 1999).

Also, a team of scientists, making measurements at Popo, a Mexican volcano, noticed similar results, measuring very high electrical charge sizes at that site, coming from the geological background (Markson & Nelson 1970).

The ability of water to produce electric charge can be seen with a Kelvin water dropper (Thomson 1872).

### *Light and Other Phenomena*

Many strange self-luminous phenomena are often manifested predominantly above seismic faults, but also above powerful geo-anomalies associated with strong metallic content without seismic record. Frequently they are manifest as colorful spheres but also they have other shapes, which are basically an indication and a result of a flowing electric current. As Dr. Levengood explains, the already electrified air molecules absorb the extra energy from the photons of a camera's flash, and are driven to an even higher energy state. Dr. Bruce Cornet, a geologist, mapped the locations of several such bright events and noticed that they were concentrated on a line of strong negative magnetic anomalies (Burke & Halberg 2005).

Also, light manifestations have been observed to pass through the ground in pre-seismic periods, due to the pressure stress of pressured rocks that emit electromagnetic signals (Kerr 1995). In the laboratory, this pressure caused similar luminous spheres, even in non-quartz rocks (Brady & Rowell 1986).

Specialists in light sphere emanations have estimated that 5kV/inch DC electric fields are enough to produce a glowing sphere of ionized air (Powell & Finkelstein 1970).

Engineer Lonetree describes the phenomenon of the non-dipolar field:

As the outer core rotates, a magnetic North and South Pole are created. During this process another form of magnetism is produced, non-dipolar in nature. This magnetism that does not have north or south. It's just pure magnetic energy. Most of this free energy is absorbed by the primary dipole,

(North/South) sector, but a portion of it can reach and penetrate the surface of our planet. Inside the earth, non-dipole magnetism takes the form of a vortex-like (spiral or circular) shape that shows up and down movement. (Miller & Lonetree 2013).

## Parameters

The parameters involved in a geophysical anomaly are the following:

- Soil and air conductivity
- DC magnetic fields
- Rotating electromagnetic fields
- Air ion intensity
- Soil temperature
- Ground and air humidity
- Seismic activity
- Infra- and ultrasounds
- Radioactivity
- Existence of groundwater
- Existence of faults
- Gravity intensity
- Geoneutrons
- Scalar waves
- Existence of quartz or magnetite in the subsoil

The strongest effects of electromagnetic forces occur on the boundary of a disturbed zone, and not at the center. On the boundary of such electrical conductivity discontinuities, extreme instabilities are observed in the vertical component of the geomagnetic field (Rikitake & Honkura 1986). The phenomenon of the intensity of these phenomena is *strengthened* at sites where the boundaries of magnetic, seismic, and gravitational zones *cross* (Burke & Halberg 2005).

## Interaction with Biology

### Effects on Humans

According to researchers, there is a disease associated with geophysical anomalies of the subsoil (Derek 1994) that affects the normal functioning of the body and can be described as a geo-pathogenic region disease (Kharat 2000).

The effect of natural electromagnetic waves and various types of geo-anomalies on human biology is obvious, due to:

- The presence of magnetite and magnetic crystals in the brain, the area of the ethmoid, and the ears (Kirshivink, Kobayashi-Kirshivink, & Woodford 1992, Barinaga 1992, Ruttan, Persinger, & Koren 1990)
- The presence of iron in the blood
- The body's composition of 70% water, which has high electrical conductivity and creates magnetic crystals (Fesenko & Gluvstein 1995)
- The property of tissues to function as semiconductors without special resistance, known as the non-thermal effect (Oschman 2000)
- The production of electromagnetic fields from the heart and the brain, the electrical transmission of signals through the nerves. The *strongest* electromagnetic field is that of the heart, *100 times higher* than that of the brain (McCraty 2003)

Humans and all living beings are, among other things, a network of the production, reception, and emission of electromagnetic fields. The electrical function of the various systems, the iron particles in the cells, the function of the proteins as semiconductors of the cell membrane components, and the intra- and extracellular water as liquid crystals, constitute biology as producer, transmitter, and receiver of electromagnetic information. It is known that cell-body microtubules are conductors of electromagnetic waves (f = 1013 Hz with their harmonics at $\lambda$ = mm) that coordinate cellular functions (Rahnama, Tuszynski, & Bókkon 2011), and that centrosomes contain silicon oxides and emit and receive electromagnetic signs.

The human body produces a series of fundamental electromagnetic frequencies that are characteristic of its structure and function (Andreev, Beliy, & Sit'ko 1984). Today, measurements of various functions can be made with modern magnetometers such as the SQUIDs, Nuclear Magnetic Resonance (NMR), and Magnetic Resonance Imaging (MRI) devices. In 1970, Russian and Ukrainian radio physicists discovered that there is resonance of tissue and cell coordination with very high frequency and low-intensity radio waves. Thus for the first time the resonant frequencies of humans, animals, and other components (biological and chemical substances) were recorded (Kositsky, Nizhelska, & Ponezha 2001). The transmission and recording of very low photon intensity activity (biophotons) from the human body, of a different frequency for each organ, is another important indication of the presence of fields (Cohen & Popp 2003).

Bioinformatics has shown that communication—the language of the body—is electrical and chemical. Nerve electrical stimuli contain information, and biochemical compounds-signals contain information to be executed, encoded within their structure. Cells are full of recipients of information (receptors) and contain all functional structures and biochemical pathways for translation and transfer of this information to the cell nuclei for execution (Oschman 2000).

Electromagnetic fields change the cell's transmissibility to ions, with an increase in calcium intake. They also alter gene expression and signal transduction inside the cells (Habash 2008).

There are two kinds of magnetic fields that have been found: The first is the one within human beings and other living beings, which is *produced* during *the transfer of ions to the nerves*, as well as the functioning of the heart and brain; the second is the Earth's own magnetic field generated from its liquid core and rotation. It seems that the two fields interact and affect the bodily activities of living beings. Becker and Selden (1985) reported that the normal geo-magnetic field plays an important role in maintaining within normal limits the direct current system of controlling bodily functions.

The SQUID magnetometer has also identified the existence of a DC peri-neural field, which produces mainly in the brain constant DC magnetic fields on the order of *one billionth* of the intensity of the geomagnetic field, which is on average about 50,000 nT. Experiments on snails have shown the dependence of biorhythms on the Earth's magnetic field.

The main process of cell division in which cell chromosomes are broken down and aligned and distributed equally between the two cells lasts only a few minutes. Several longer stages are needed, one of which is the duplication of all cellular DNA. All stages together last one day. Therefore, cell growth and repair, which is based on the regulation of cell division, is synchronized with the Earth's magnetic field (Becker & Selden 1985).

Becker and Selden (1985) made an experiment to observe the dependence of the main human biorhythms on the geomagnetic field. They isolated two groups of people in two underground rooms, one of which was blocked from any activity indicative of the passage of time, and the other blocked also from the geomagnetic field. He found that in both rooms there was a disturbance of the biorhythms, which was translated as an extension of the rhythms in the room that had been blocked from the geomagnetic field. When volunteers in this room were exposed to a 10 Hz frequency range (0.025 V/cm), similar to Earth, the rate disorder was restored (Becker & Selden 1985).

Detailed studies show that all vertebrates have a similar magnetic instrument in the area of the ethmoid, and this instrument transmits biorhythmic time elements from the microarrays of the geomagnetic field to the pineal gland (Becker & Selden 1985).

Scientists in India decided to measure the effect of electromagnetic fields of different frequencies on humans (Subrahmanyam, Narayan, &

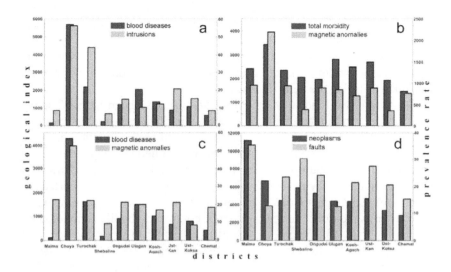

**Figure 2. Relationships between prevalence rates in the adult population and geological indices in 2002.**
(a) blood diseases versus intrusions
(b) total morbidity versus magnetic anomalies
(c) blood diseases versus magnetic anomalies
(d) neoplasms versus faults (Shitov 2010)

Srinivasan 1985). The parameters recorded were cardiac pulse, pressure, brain waves, and blood levels of neurotransmitters. They noticed that volunteers had the most reactions at frequencies of 0.01 Hz in direct current and 50 nT in a produced magnetic field. The above are characteristics of daytime variations of the geomagnetic field and especially in an area with anomalies (Burke & Halberg 2005).

Shitov (2010) found a very strong correlation among blood, nervous, respiratory, and genito–urinary system diseases and geophysical anomalies (Figure 2).

Persinger and Psych (1995) studied geo-anomalies' influences in connection with investigations related to unexplained deaths in epilepsy patients that are maximized at sunrise as the fluctuations of the geomagnetic field peak just before sunrise. Michael Persinger stated:

> Temporal and regional variations in psychological processes have been associated with three geological factors: geochemical features, geomagnetic changes, and tectonic stress. In the geochemical field, the presence

of copper, aluminum, zinc, and lithium can affect the incidence of thinking disorders, such as schizophrenia and senile dementia. These common elements are found in many soils and groundwater.

Geomagnetic abnormalities have been associated with increased anxiety, sleep disturbances, altered mood, and a higher incidence of psychiatric admissions. Transient and local epidemics of strange and unusual behaviors are sociological phenomena, which have increased before seismic activity has increased in a region, and are associated with tectonic strain.

Many of the modern associations between geological parameters and human behavior are evident in historical data. The effects of geophysical and geochemical parameters on human behavior are often complex and are not detected by the limited scope of most studies. (Persinger 1987)

He also concludes that cerebral function associated with consciousness responds to subtle changes in geomagnetic activity. Measuring the results of the same changes in his laboratory, he noticed that they had a direct effect on the electrical sensitivity of brain cells (and in extreme form caused seizures) and influenced the ability of individuals to concentrate during the day. He chose in his experiments to reproduce the magnetic field variation of 50 nT that is consistent with the fluctuations usually occurring in nature.

The above confirms that all human beings are permeated with the geomagnetic field, and we are connected to it, as well as all the secondary fields resulting from this connection. As a result, very small changes in geomagnetic field activity directly affect human biology, and create the ability to change one's brain without humans realizing it (Persinger & Levesque 1983).

In the research done by Rudnik and Melnikov (2010), it was clearly shown that cancer incidence was greatly augmented in the intersection of fault lines (Figure 3).

Geophysicist Andrei Apostol used his own device to measure the number of muscle contractions in volunteers as they moved into different geological backgrounds. The results showed a strong correlation among muscle contractions, gravitational anomalies, and geologic incisions (Apostol 1995).

A powerful example of a geo-anomaly is the so-called Cliff of Tears, an area in North America where visitor feedback revealed that male visitors would nosebleed, while women had sudden menstruation. David Barron, director of Gungywamp Swamp in Connecticut, conducted an experiment with 20 volunteers and nurses who recorded a significant difference in blood pressure after exposure to the area (Burke & Halberg 2005).

In a research experiment done on more than 800 people, called the Vienna Report, a group of 20 scientists found significant results on people

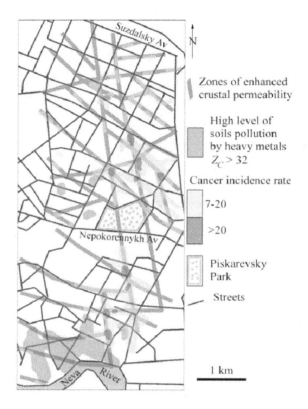

**Figure 3. Portion of the Kalininsky District of Saint Petersburg: Cancer incidence rate (per 1,000 population) in the years 1991–1992, ZEPC, and areas marked by a high level of soil pollution with heavy metals (Rudnik & Melnikov 2010).**

exposed to a geophysical anomaly. There were significant changes in serotonin, zinc, and calcium levels, and also in immunoglobulins such as IgA (Bergsmann 1990).

Dubrov (2008) studied areas with geophysical anomalies and found that there was much higher morbidity regarding various diseases in comparison with areas having homogeneous geophysical parameters (Table 1).

Dharmadhikari defines *geopathic zones* as places on earth known for causing *health problems*. He and his team made measurements in individuals in a geopathic and a neutral zone. The results showed that the

**TABLE 1**

**Incidence of Various Diseases in People Living on a Geophysical Anomaly and in an Undisturbed Zone (Dubrov 2008)**

| Disease | Geophysical Anomaly | Undisturbed Zone |
|---|---|---|
| Total | $1205 \pm 25$ | $792 \pm 5$ |
| Infections | $45.4 \pm 0.3$ | $17.7 \pm 1.9$ |
| Oncology | $7.96 \pm 0.15$ | $5.58 \pm 0.04$ |
| Mental | $4.94 \pm 0.08$ | $1.14 \pm 0.23$ |
| Hypertension | $4.48 \pm 0.17$ | $0.83 \pm 0.02$ |

electrical potential of the body increased and skin resistance was reduced when they were exposed to a geopathic zone in comparison with the neutral zone (Dharmadhikari et al. 2011).

Aschoff (2014), a physician, was the first to use the blood's electromagnetic oscillations, which are measurable by a simple blood test (Aschoff et al. 1994). After 20,000 tests, he noticed that people with electromagnetically oscillating blood lived *without exception* in a disrupted geopathic zone, either in their sleeping area or in their workplace. Individuals with only magnetically oscillating blood were not exposed to a geopathic disorder and were healthy. Due to the stress from the electric current and radiation (mostly gamma ray) emitted by a geopathic zone, the blood loses its natural structure and becomes electrically polarized in the opposite charge. Aschoff also mentions that neutron radiation emitted by geopathic zones can cause mutations in the cells.

Hacker, also a medical doctor, reports that longitudinal scalar waves emitted at different locations can cause various effects in the form of symptoms on biological systems. Together with his team, he conducted experiments using Dr. Korotkov's GDV (Gas Discharge Visualization), while making Immunoglobulin-A (IgA) and A-Amylase measurements. The results were in complete harmony; as in the higher GDV mean area, IgA levels were also higher, indicative of relaxation; and in the geopathic zones with lower GDV Mean Area, A-amylase was higher, indicative of a stress state. Corona Discharge diagrams of GDV showed weakening of the immune system and epiphysis function in the case of exposure to the geopathic zones (Hacker, Augner, & Pauser 2011) (Figure 4).

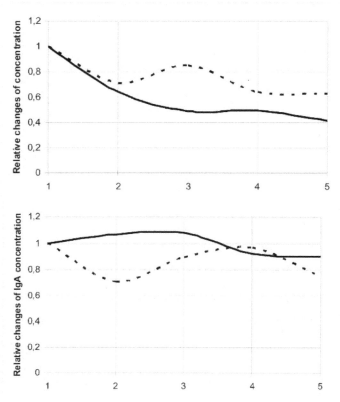

**Figure 4. TOP:** The normalized time course (9:00 a.m. to 1:00 p.m.) of the cortisol concentration in saliva obtained at a potential "neutral zone" (black line), and at a potential "geopathic stress zone" (dotted line). Note that the stress zone in times led to increased cortisol levels compared with those obtained from the more neutral place.
**BOTTOM:** The normalized time course (9:00 a.m. to 1:00 p.m.) of salivary IgA measured at a potential "neutral zone" (black line), and at a potential "geopathic stress zone" (dotted line). The potential stress zone gave a slightly different time course to that obtained at the more neutral place (Hacker et al. 2008).

A secondary effect of telluric currents is that they attract airborne electrically charged particles of the opposite charge, which can have a significant effect on biology. The effects of breathing air ions have been shown in detailed studies. In modern buildings with enhanced crust permeability (such as faults), a decreased number of negative ions has been observed (Rudnik & Melnikov 2010), which can adversely affect human health, suppressing the immune system (Krueger 1972, Tchijevsky 1929).

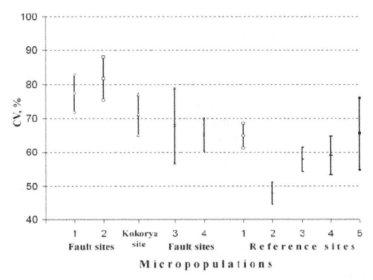

**Figure 5. Variation (CV) of the fruit shape at a fault and at neutral sites (Boyarskikh & Shitov 2010).**

### *Flora and Fauna*

Obviously, the geopathic zones affect not only humans but also all species of animals, plants, fungi, and bacteria (Dubrov 2008, Gak & Gridin 2008, Hacker et al. 2008, Von Pohl 1983).

One ancient method for choosing a residence location was noticing the presence of shrubs, land color, presence of water systems, and tree growth. Also, another method used was housing animals in the proposed area and observing their behavior and health for a certain period of time (Bradna 2002). Thus it is obvious that the effect of earth radiation anomalies on plants and animals has been known for many thousands of years.

Tombarkiewicz (1996) in her research proved the effect of geomagnetic anomalies on cows' health. When they were moved to a location with anomalies, they developed health problems above a normally expected ratio, and also showed lower levels of zinc, copper, and iron.

Yeagley (1947) showed that pigeons have a magnetic sensation that allows them to use the geomagnetic field as a compass. Magnetic crystals have been found in almost all animals using the Earth's magnetic field for navigation (Long 1991). Regarding plants, Boyarskikh and Shitov (2010) showed that fruit plants located within geophysically anomalous areas produced smaller fruits and increased diversity of fruit (Figure 5), and an increase in the expression of recessive traits.

Plant intrapopulation variability increase occurs in the least favorable environments according to Mamaev (1973). Geophysical anomalies act as stress parameters for plants.

## Conclusion

Various parameters constitute a geophysical anomaly. Among them are intense change in the magnetic and/or gravity field, change in radiation or radioactivity levels, conductivity discontinuity of the ground material, the presence of a fault, and/or subterranean water.

The effect of these anomalies on human health can be explained through the presence of magnetite, iron, water, and also the body's own electromagnetic fields (brain, heart, neural network). Adverse effects are also shown in flora and fauna.

## Acknowledgments

Special thanks to IKY, the State Scholarship foundation in Greece, for financial support of this doctoral thesis research.

## References Cited

Adler, M. P., Le Mouel, J. L., & Zlotnicki, J. (1999). Electrokinetic and magnetic fields generated by flow through a fractured zone: A sensitivity study for La Fournaise volcano. *Geophysical Research Letters, 26*(6):795–798, March 15.

Andreev, Y. A., Beliy, M. U., & Sit'ko, S. P. (1984). Occurrence of the human body's own characteristic frequencies. *Documents of AN USSR*, pp. 60–63.

Apostol, A. (1995). North American Indian effigy mounds: An enigma at the frontier of archaeology and geology. *Journal of Scientific Exploration, 9*(4):549–563.

Aschoff, D. (2014). *Electromagnetic Property of Blood Measurable Differences on Geological Fault Zones—Detection of Fault Zones by Measurement of a Drop of Blood* [original in German: Elektromagnetische Eigenschaft des Blutes durch Reizzonen messbar verändert—Feststellung von Reizzonen Einwirkung durch einen Tropfen Blut]. In *Der Elektromagnetische Bluttest, Yashi Kunz*. Lightball Media.

Aschoff, D., Dräger, M., Heinz Müller, H., Aschoff, J., Gruner, S., Rothdach, P., Schumacher, E., Uiblacker, K., & Worsch, E. (1994). *Location as a Risk Factor: Mutation and Measurement of Biologically Effective Radiation and Fields and Their Influence on Humans* [original in German: Standort als Risikofaktor: Mutung und Messung biologisch wirksamer Strahlen und Felder und ihr Einfluss auf den Mensche]. Reich Verlag.

Barinaga, M. (1992). Giving personal magnetism a whole new meaning. *Science, 256*:967.

Becker, R., & Seldn, G. (1985). *The Body Electric*. New York: William Morrow, pp. 240, 245.

Bergsmann, O. (1990). *Risikofaktor Standort*. Vienna: Facultas University Press.

Boyarskikh, I. G., & Shitov, A. V. (2010). Intraspecific Variability of Plants: The Impact of Active Local Faults. Chapter 5 in *Man and the Geosphere* edited by I. V. Florinsky, Happauge, NY: Nova Science Publishers.

Bradna, L. (2002). *The Influence of Hydro Pathogenic Zones on Drivers*. Pune, India: Narendra Prakashan, pp. 38–43.

Brady, B. T., & Rowell, G. A. (1986). Laboratory investigation of the electrodynamics of rock fracture. *Nature, 321*:490.

Burke, J., & Halberg, K. (2005). *Seed of Knowledge, Stone of Plenty: Understanding the Lost Technology of the Ancient Megalith-Builders.* San Francisco: Council Oak Books.

Cohen, S., & Popp, F. A. (2003). Biophoton emission of human body. *Indian Journal of Experimental Biology, 41*:440–445.

Croome, D. J. (1994). The effect of geopathic stress on building occupants. *Renewable Energy, 5*(58):993–996.

Cull, J. P., & Tucker, D. H. (1986). Telluric currents and magnetic anomalies. *Geophysical Research Letters, 13*(9):941–944, September.

Dharmadhikari, N. P., Meshram, D. C., Kulkarni, S. D., Kharat, A. G., & Pimplikar, S. S. (2011). Effect of geopathic stress zones on human body voltage and skin resistance. *Journal of Engineering and Technology Research, 3*(8):255–263, August.

Dubrov, A. P. (2008). Geopathic Zones and Oncological Diseases. *Proceedings of the Sixteenth BDA Congress on "Earth's Fields and Their Influence on Human Beings"*, Druskininkai, Lithuania, June, pp. 42–44.

Fesenko, E. E., & Gluvstein, A. Y. (1995). Changes in the state of water, induced by radiofrequency electromagnetic fields. *FEBS Letters, 367*:53–55.

Florinsky, I. V. (Editor) (2010). *Man and the Geosphere.* New York: Nova Science Publishers.

Freund, F. T. (2003). Rocks that crackle and sparkle and glow: Strange pre-earthquake phenomena. *Journal of Scientific Exploration, 17*(1):37–71,

Gak, E. Z., & Gridin, V. I. (2008). About Nature's Influence on Geophysics and Earth's Fields Anomalies on Living Systems. *Proceedings of the Sixteenth BDA Congress "Earth's Fields and Their Influence on Human Beings"*, Druskininkai, Lithuania.

Habash, R. W. (2008). *Bioeffects and Therapeutic Applications of Electromagnetic Energy.* Boca Raton, FL: CRC Press.

Hacker, G. W., Augner, C., & Pauser, G. (2011). Daytime related rhythmicity of gas discharge visualization (GDV) parameter glow image area: Time course and comparison to biochemical parameters measured in saliva. In *Energy Fields: Electrophotonic Analysis in Humans and Nature* edited by Korotkov, St. Petersburg, Russia, pp. 214–232

Hacker, G., Pauser, G, & Augner, A. (2011). Geophysical background, target structures, and effects of geopathic stress zones, as detected with gas discharge visualization (GDV) methodology. In *Spiral Traverse, Journey into the Unknown* edited by K. Korotkov, pp. 315–348.

Hacker, G. W., Eder A., Augner, C., & Pauser, G. (2008). Geopathic Stress Zones and Their Influence on the Human Organism. *Proceedings of the Sixteenth BDA Congress on "Earth's Fields and Their Influence on Human Beings,"* Druskininskai, Lithuania, 21 pages.

Hessler, V. P., & Wescott, E. M. (1959). Correlation between earth-current and geomagnetic disturbance. *Nature, 184*:627.

Hippocrates [ππokράτης] (1992). *Concerning Nutrition [Περί Διαίτης].* Odisseas Chatzopoulos.

Kerr, R. A. (1995). Quake prediction tool gains ground. *Science, 270*(5238):911–912

Kharat, A. G. (2000). Theoretical and Empirical Investigations of the Built Environment. Ph.D. thesis, Pune University, India.

Kirshivink, J. L., Kobayashi-Kirshivink, A., & Woodford, B. J. (1992). Magnetite biomineralization in the human brain. *Proceedings of the National Academy of Science, 89*:7683–7687.

Kositsky, N. N., Nizhelska, A. I., & Ponezha, G. V. (2001). Influence of high-frequency electromagnetic radiation at non-thermal intensities on the human body. *No Place to Hide Newsletter of the Cellular Phone Taskforce Inc., 3*(1):1–33.

Krueger, A. P. (1972). Are air ions biologically significant? A review of a controversial subject. *International Journal of Biometeorology, 16*(4)(December):313–322.

Langer, H. D. (1997). Das geophysikalische Standortproblem der Solitärbäume, Teil 1: Ergebnisse systematischer Naturbeobachtungen [The geophysical location problem of solitary trees, Part 1: Results of systematic observations of nature]. *Veröffentlichungen des Museums für Naturkunde Chemnitz, 20*:115.

Langer, H. D. (2008). A first consistent physical model of radiestesy? *Proceedings of the Sixteenth BDA "Earth's Fields and Their Influence on Human Beings"*, Druskininskai, Lithuania, June.

Miller, I., & Lonetree, B. (2013). The Sedona Effect: Correlations between geomagnetic anomalies, EEG brainwaves & Schumann Resonances. *Journal of Consciousness Exploration & Research, 4*(6).

Long, M. (1991). Secrets of Animal Navigation. *National Geographic Magazine*, pp. 70–99.

Mamaev, S. A. (1973). *Forms of Intraspecific Variability in Woody Plants*. Moscow: Nakua. [in Russian]

Mamirova, G. N. (2010). Bioindication of Geological Anomalies in Ecosystems. Dissertation on Biological Science, Republic of Kazakhstan, City of Almaty.

Markson, R., & Nelson, R. (1970). Mountain-peak potential-gradient measurements and the Andes glow. *Weather, 25*(8):350–360.

Martin, R. J., III, Haupt, R. W., & Greenfield, R. J. (1982). The effect of fluid flow on the magnetic field in low porosity crystalline rock. *Geophysical Research Letters, 9*(12):1301–1304.

McCraty, R. (2003). The Energetic Heart: Bioelectromagnetic Interactions Within and Between People. Heartmath Institute.

Mizutani, H., Ishido, T., Yokokura, T., & Ohnishi, S. (1976). Electrokinetic phenomena associated with earthquakes. *Geophysical Research Letters, 3*(7):365–368.

Oschman, J. L. (2000). *Energy Medicine: The Scientific Basis*. Philadelphia, PA: Churchill Livingstone.

Persinger, M. A. (1987). Geopsychology and geopsychopathology: Mental processes and disorders associated with geochemical and geophysical factors. *Experientia, 43*:92–104.

Persinger, M. A., & Levesque, B. F. (1983). Geophysical variables and human behavior: XII. The weather matrix accommodates large portions of variance of measured daily mood. *Perceptual and Motor Skills, 57*:868–870.

Persinger, M. A., & Psych, C. (1995). Sudden unexpected death in epileptics following sudden, intense, increases in geomagnetic activity: prevalence of effect and potential mechanisms. *International Journal of Biometeorology, 38*(4):180.

Powell, J. R., & Finkelstein, D. (1970). Ball Lightning: Less well known than stroke lightning, ball lightning is about as frequent and can be simulated in the laboratory. *American Scientist, 58*(3):272.

Rahnama, M., Tuszynski, J. A., & Bókkon, I. (2011). Emission of mitochondrial biophotons and their effect on electrical activity of membranes via microtubules. *Journal of Integrative Neuroscience, 10*(1):65–68.

Rikitake, T., & Honkura, Y. (1986). *Solid Earth Geomagnetism*. Tokyo: Scientific Publishing, pp. 296–325.

Rudnik, V. A., & Melnikov, E. K. (2010). Pathogenic Effect of Fault Zones in the Urban Environment. In *Man and the Geosphere* edited by I. V. Florinsky, New York: Nova Science Publishers, pp. 169–183.

Ruttan, L. A., Persinger, M. A., & Koren, S. (1990). Enhancement of temporal lobe-related experiences during brief exposures to milligaus intensity extremely low intensity magnetic fields. *Journal of Bioelectricity, 9*(1):33–54.

Shitov, A. V. (2006). Ecological consequences of activization of geodynamic processes in the Mountain Altai. Vestnik Tomskogo Universiteta, *Bulletin Operativnoi Nauchnoi Informatsii, 72*:118–135 [in Russian, with English abstract].

Shitov, A. V. (2010). Health of people living in a seismically active region. Chapter 7 of *Man and the Geosphere* edited by I. V. Florinsky. New York: Nova Science Publishers.

Subrahmanyam, S., Sanker Narayan, P. V., & Srinivasan, T. M. (1985). Effect of magnetic micropulsations on the biological systems—A bioenvironmental study. *International Journal of Biometerology, 29*(3):293–305.

Tchijevsky, A. L. (1929). L'aéroionothérapie des maladies pulmonaires. Recherches expérimentales de l'effet de l'air ionisé sur la tuberculose des poumons, la pneumonie et la bronchite. *La Presse Thermale et Climatique, 70*:653–665.

Thomson, W. (1872). *Reprints of Papers on Electrostatics and Magnetism*. London: Macmillan.

Tombarkiewicz, B. (1996). Geomagnetics studies in cow house. *Acta Agriculturae Scandinavia Section A, 426*:34.

United Nations Scientific Committee on the Effects of Atomic Radiation (UNSCEAR) (1993). Sources and Effects of Ionizing Radiation: UNSCEAR 1993 Report to the General Assembly, with Scientific Annexes. United Nations Publications.

Von Pohl, F. G. (1983). *Earth Currents as Pathogenic Agents for Illness and the Development of Cancer*. Freich Verlag. [out of print]

Yang, J., Lu, F., Kostiuk, L. W., & Kwok, D. Y. (2003). Electrokinetic microchannel battery by means of electrokinetic and microfluidic phenomena. *Journal of Micromechanics and Microengineering, 13*(6)(October 20):963–970.

Yeagley, H. (1947). A preliminary study of a physical basis of bird navigation. *Journal of Applied Physics, 18*(12):1035–1063.

*Journal of Scientific Exploration*, Vol. 32, No. 3, pp. 514–554, 2018      0892-3310/18

*RESEARCH ARTICLE*

# A Methodology Proposal for Conducting a Macro-PK Test on Light Spinning Objects, in a Non-Confined Environment

### Eric Dullin

ericdullin@gmail.com

### David Jamet

Psychophysics Anomalies and Cognitive Dissonance Laboratory (LAPDC), Poitiers, France

Submitted November 29, 2017; Accepted July 21, 2018; Published September 30, 2018

DOI: https://doi.org/10.31275/2018.1266

**Abstract**—For more than a century, there has been much debate around the use of telekinesis–psychokinesis (TK)[1] to explain the rotating movement of light objects on an upright standing needle in the proximity of a hand. Thermally induced aerodynamic effects have been considered as likely physical explanation factors. Despite this controversy, many people still upload videos claiming the phenomenon on the Internet. Most of the scientific studies performed were focused on whether or not the effects could be observed if some physical constraints were added, in order to avoid the aerodynamics factor, or if the same results could be reproduced using some thermic or/and aerodynamic artefacts instead of a human presence. The first approach runs the risk of inhibiting a phenomenon about which little is known. The second has not yet shown clear reproducible experiences, which produce the same results as with a human presence, except in very specific situations. Our objective is to be able to detect and study psychokinesis in confined and non-confined environments with scientific measurement tools. Our hypothesis is that there could be a way to separate psychokinetic effects from aerodynamic effects, even in a non-confined environment, thereby avoiding the drawbacks of the first approach. This technique of approaching anomalous perturbations could be described as partial physical isolation of the target system, with a measurement system ensuring control of the remaining known effects. It can be related to two other techniques previously described (May, Utts, & Spottiswoode 1995: Introduction). From the beginning, the LAPDC (Psychophysics Anomalies and Cognitive Dissonance Laboratory) has been fostering a PKers (subjects practicing psychokinesis) volunteer team in order to do the experiments. From 2012 to 2016, we were developing specific scientific methods in order to study the psychokinesis effect on a spinning mobile with or without confinement. More specifically, we developed a protocol starting with particle image velocimetry (PIV) in order to measure the air-flow speeds around the mobile. Further research has driven us to create a set of processes using MATLAB, which we named Scan-Flow-Mobile. It has enabled us to construct one global model, integrating air flow movements and mobile movements, and to scrutinize it. Using this, we were able to compare different experiments. We conducted a thorough analysis of the interaction between the mobile and the air flows, and studied the cause-and-effect relationships between their movements. A review of the "spinning mobiles" literature of the last century, either with a psychokinetic hypothesis or an aerodynamic/thermic explanation, has been done. We also studied other potential causes of motion such as electrostatic forces, magnetism, vibrations, and the impact of radiation. Then, as a pilot study, we conducted eight experiments in non-confined environments, with three setup categories: one where the mobile motion was driven by generated air flows (A), one in which a motor drove the mobile (M), and the last one where a PKer drove the mobile (PK). The ratio (mobile speed / mobile periphery airflow speed) was used as a way to compare experiences between experiments and categories. In this paper we focus only on the methodological approach and so on the categories A and M. With regard to this ratio, the category M experiments stayed above or equal to 2, while category A was below or equal to 0.5. This clearly separated purely aerodynamic effects (A) from the motor-driven effect (M). The methodology could be a good candidate to conduct a macro-PK test in a non-confined environment with the capability to eliminate or not the aerodynamic effect as the explanation. A potential bias and errors analysis is presented, which takes into account the difference between air-flow and smoke-particle velocity, the mean speed evaluation for the air flows and the mobile. Indeed, we evaluated the potential error on the ratio air-flow speed/mobile speed as approximately ±8.9%, which is marginal compared with the ratios differences between categories A and M. The methodology also presents some features that help to detect tricks that could be tried by some misbehaving PKers (e.g., mouth air blowing, hand movements, etc.). We will continue to look to improve documentation of the total measurement process, in order to give other laboratories the potential to test it in their experiments.

## Introduction

### *Is Psychokinesis on Small Objects Widespread or Non-Existent?*

Nowadays, we can find on the Internet many videos with different kinds of "psi wheels" claiming the evidence of a psychokinetic effect.[2] Many Internet forums[3] are skeptical about these claims, however, and announce that all of these videos are fakes, or at least show some classical aerodynamic effect induced by the heat of the hand.

Where does the truth lie? Are there scientific studies available on the subject?

It is well-known that it is not possible to prove any null hypothesis. The null hypothesis is generally assumed to be true until evidence indicates otherwise. Here, the null hypothesis is: "There is no relationship between a 'psi wheel' movement and a human presence, except through known physical forces."

### A Long Story from Mesmerism and Magnetic Fluid to Vital Energy

By the end of the 18th century, Franz-Anton Mesmer had developed his animal magnetism theory, proposing that all living beings were emanating some kind of "fluid" that was capable of healing others and inducing trances. This theory had great influence until the beginning of the 20th century, and several individuals tried to develop a mechanism to measure this ostensible "magnetic fluid" or "bio fluid."

The French physician Hypolyte Baraduc (1893) used an instrument he called a biometer, which consisted of a glass jar inside of which was a needle suspended by a thread. The needle rotated on top of a circular surface with numbers, which allowed Baraduc to obtain readings corresponding to the movement of the needle. When people put their hands near the instrument, the right hand effected an attraction on the needle whereas the left hand repelled it. These movements, described by Baraduc as "tangible and recordable expressions of a superior Force," were considered to be evidence for the existence of a vital body in human beings (Alvarado 2006).

Another French parapsychologist, Paul Joire, devised an instrument call a sthenometer (Figure I1). This, he claimed, could detect the "nervous force" emitted by the body, and he published several papers on the topic.

Le Comte de Tromelin, a French physician and inventor, proposed a motor driven by "human fluid," which was easy to build (Tromelin 1907). He produced several publications such as the one in *Le Monde Psychique* (Tromelin 1912) from whom the pictures below are extracted (Figure I2). These motors were mainly composed of a light object attached to a needle which sat on top of a metallic support (or at least a smooth surface so that the needle could spin.

People then performed more tests and research on these kinds of mobiles. Pierre Archat built some testing equipment using an approach to prevent or balance the aerodynamic thermally induced forces, in order to see if some remaining effects persisted (Archat 1908). His objective was to study whether any person could generate a motor action on a living being or on an object in their proximity. Figure I3 shows one of the setups he used. Here, the object O is placed on a torsion mobile B, itself inside a bell jar C. The subject approaches it with their hand, M, close to C without touching it.

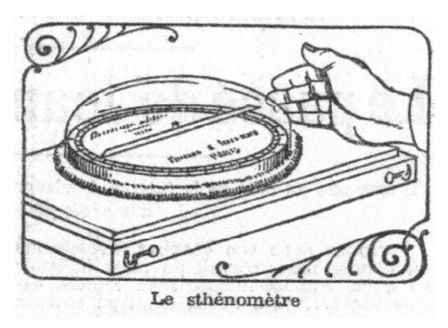

Le sthénomètre

**Figure I1. In** *Le Petit Parisien* **newspaper, December 26, 1908.**

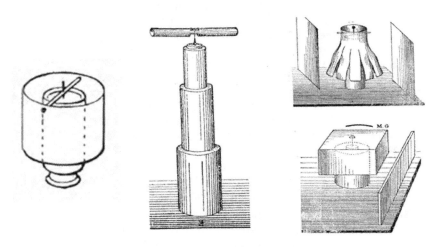

**Figure I2. Examples of Tromelin's motors (Tromelin 1912).**

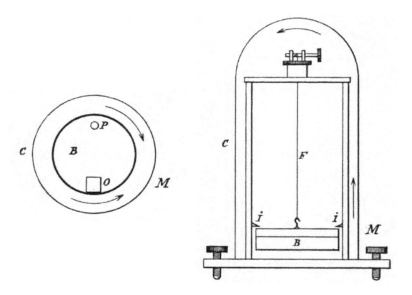

**Figure 13. Archat "telekinesis" testing machine (Archat 1908).**

Here is an extract of his translated conclusion (Archat 1908):

It is right that this apparatus is less sensitive than the first one because the weight of B is ten times the one with only a needle. But no deviation, even small, has been observed, even with the use of a 5–6 times magnifying glass. My research stopped there. If the result is negative, I think it's not a good approach to conclude that the force I was looking for didn't exist, but we need to study it by other means than the one I used. It seems that we could deduce from these experiences that this force is not emitted in a continuous way and with an appreciable strength by the human organism. Perhaps it exists in a latent state in the organism and manifests itself only in certain conditions. Indeed, it is possible that the result could have been different with a medium with physical effect: This experiment has not been done.

At the same Congress, René Warcollier outlined his analysis and criticisms of Tromelin's book (Warcollier 1908). He proposed some explanations as described below (Figure 14): Human body heat creates a rising airstream. This upward airflow draws air above the table, and induces an airstream in the direction of the subject. The hand, forming an obstacle to this airstream on the right, creates a dissymmetry in the aerodynamic forces

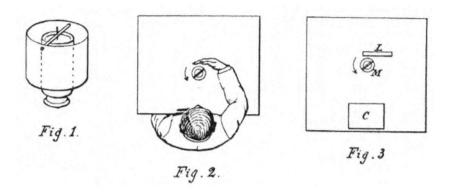

**Figure I4. Warcollier's explanation of Tromelin's motor movement (Warcollier 1908).**

around the mobile. This creates a resulting torque that spins the mobile. Figure 3 inside Figure I4 proposed a way to simulate the phenomenon without a human presence using a calorimeter as a heat source and a book as an obstacle.

In Russia, Yakov Perelman published the first release of his book in Russian *Physics for Entertainment* (Perelman 1913), translated later into many languages. In his book, on p. 117, he exposed "The mysterious twirl" with the picture in Figure I5. His explanation was:

> when you bring your hand up, the air near it, which is warmed by its proximity, rises, and, pressing against the piece of paper, causes it to spin. It revolves because it is slightly folded, thus performing the same role as a curled piece of paper suspended above a lamp.

Then he referenced Nechayev, who in 1816 wrote a communication to the Moscow Medical Society entitled "The Gyration of Light Bodies Caused by the Heat of the Hand" (we did not succeed in obtaining this communication). Using this reference, Perelman explained why the piece of paper always gyrates in one and the same direction—from the wrist toward the fingertips. It is because the fingertips are always colder than the palm of the hand; consequently, the palm gives rise to a stronger ascending air current than the fingertips do.

In France, Clément Martin did a further analysis of the Archat/Warcollier results using an apparatus on which people can apply the hand touching it at different positions (Martin 1926). With two glass containers V and V', it was able to simulate the temperature difference between the center of the

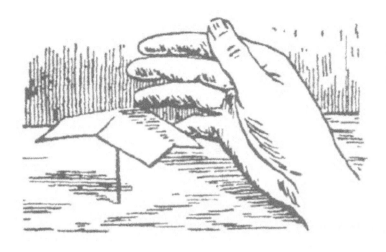

Fig. 82. Why does this piece of paper spin?

**Figure I5. The mysterious twirl (Perelman 1913).**

hand and the fingertips (Figure I6). His final statement was that no force other than an aerodynamic and thermic one has been demonstrated, and that in a confined environment serious consideration must to be given to the size of the "air column" (Martin 1926).

Tromelin's motor came back in another form with the work of the Czech physicist Julius Krmessky, who developed some mobiles with light objects which were very close to Tromelin's (Figure I7). He used them as "ideal research tools, since they are simple, inexpensive, and require no special training or psychic talent for their operation." Krmessky recommends isolation of the system from the motion of air and the effects of heat radiation by enclosing it in glass, metal, or other containers, with provision for inspection through a glass cover. Here are some of Krmessky comments about his results:

> Movement in such enclosed spaces are slow and hence not too spectacular, but are nevertheless convincing. The slow rate of the motion or the occasional immobility cannot be explained by the walls being impenetrable to outside impulses, because the device is able to detect the nearing of a hand, even through a thick layer of lumber, metal, water, etc. The cause lies somewhere else. (Krmessky 1975)

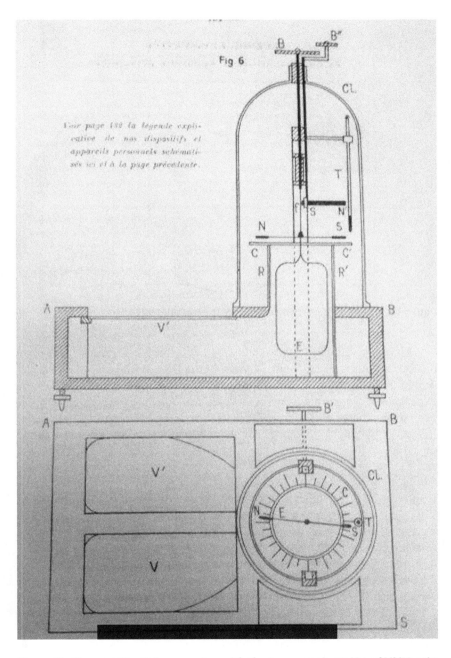

**Figure l6.  Clement Martin's apparatus with the two containers V and V' (Martin 1926).**

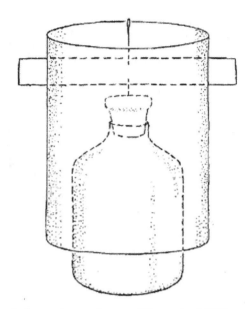

**Figure I7. Krmessky's pyschotronic rotor (Krmessky 1975).**

Thanks to Edwin C. May and Loyd Auerbach (Auerbach 1996), we discovered the work of Martin Caidin in the USA. Following the publication of G. Harry Stine's book (Stine 1985), Martin Caidin conducted many experiments around the "Energy-Wheel" (an object, like a pyramid of paper, balanced on a pin) and wrote his journal of the experiment "The Merlin Effect," in which he advocated for a specific effect with hands close to the object (even inside a jar) but also from a distance (by using a mirror to look at the target or even through a TV Monitor or a camera). However, he pointed out the importance of the distance on the strength of the effect. He also reported being able to control in some cases in which direction the wheel should spin, to stop it, and to reverse the sense of rotation. He also observed a growing difficulty in increasing the weight of the target (he experienced a learning curve from a tenth of a gram to 450 g). He also mentioned the "use it or lose it" rule: If you do not practice for a while, you have to start from the beginning with a low weight and climb the ladder again with increasing weight targets.

Thanks to Professor Peter Mulacz (head of the parapsychology association in Austria), we discovered the work of Albert Hofmann (Hofmann 1919/1992). We did not succeed in getting a copy of the book (the title can be translated as "the mystery of the radiating hand"), but Professor Mulacz summarizes Hofmann's finding as:

**Figure I8. The Egely wheel
(Egely 2017).**

* with a hollow rubber hand filled with 37° C water ==> no effect
* with a pulsation of the adjacent air ==> mobile starts rotating
* his interpretation: It is the pulsation of the bloodstream in the hand
  (at the wrist, where the doctor feels the pulse) that is the cause of
  that air effect.

More recently, John Rudkin published an article focused on the Egely wheel. This instrument is a commercial product (Figure I8) which the manufacturer claims is able to measure life energy, chi, or vitality. The subject just has to look at the speed of the wheel when their hand is placed close to it (Egely 2017). It is a typical "psi wheel": a light plastic wheel of approximately 3.5 inches spinning on a pin. Rudkin's concluding statement on his study was "The behavior I have observed of the Egely wheel and similar spinners is explicable in terms of thermally induced aerodynamic effects" (Rudkin 2001).

Although it is not a spinning mobile on a needle, but rather a torsion pendulum with small effect, we could also mention:

— The work of J. N. Hansen and J. A. Lieberman on a torsion pendulum placed as a helmet above the subject's head to detect brain impact without contact (Hansen & Lieberman 2013).

— The research of Dr. Antonio Giuditta on the human bioenergy field, with a torsion pendulum placed at a distance of one meter from the subject (Giuditta 2014).

Finally we would like to mention the recent work of John G. Kruth, who led experimental sessions with an exceptional participant using an Egely wheel (Kruth 2016). The Egely wheel was covered by a plastic container sealed with the support. The protocol eliminated other causes of movement. The participant was able to move the wheel at a distance of up to 12 inches. On average, a full rotation required three minutes. The participant required time (one day) to adapt herself to the laboratory's Egely wheel, even if she had her own Egely wheel.

## TK–PK

In Varvoglis and Bancel (2015), psychokinesis (PK) is defined as follows: "PK is the putative ability of organisms to affect other systems—both animate and inanimate—without mediation of any known physical forces or energies." We agree with this neutral definition, in which the mind/human intention is not necessarily concerned. So, at the LAPDC we prefer to use the term telekinesis (TK) instead of PK. However, as the vast majority of parapsychological publications in English explicitly use the term PK, we use the term PK in this paper (and PKers for persons practicing psychokinesis).

## Macro-PK or Micro-PK

As described in Watkins (2015), macro-PK is a term used to describe ostensible psychokinetic effects so strong or so dramatic as to require no use of statistics. It can also include levitation and teleportation. Even with small effects, we do not need to use statistics to study the phenomenon described above. Therefore, we are in the macro-PK space, also named macro anomalous perturbation (macro-AP) (May, Utts, & Spottiswoode 1995).

## Our Objective: A Third Way to Study Anomalous Perturbations

For most of the previous research described above, the pros and cons of the existence of small psychokinetic effects, producible by non-exceptional subjects, are presented. They also describe, as presented by Jahn and Dunne (2011), that uncertainty using fewer constraints could be a prerequisite for seeing larger phenomena. Indeed, as we saw above, the effect seems to disappear or become very small when increasing constraints and confinement are applied (for example, Krmessky's low motor speed in the confined environment and the Egely wheel speed in Kruth's experiment).

Our objective is to study these small effects in more detail, with new scientific means and methods that do not require confinement, in order to identify whether aerodynamic forces could really explain all these mobile-spinning effects or not. In fact, our proposition is an extension of that by May, Utts, and Spottiswoode (1995:p.196) that two techniques have been employed to study anomalous perturbations. These are:

1. complete physical isolation of the target system,
2. counterbalanced control and effort periods.

In this paper we try to show that there could be a third way: "partial physical isolation of the target system, with a measurement system ensuring control of the remaining known effect." The advantages are developed in the Methods section, in the subsection entitled ***Looking for Clear Anomalies versus Significant Deviation from Null Hypothesis on a Set of Calibrated Experiences***.

## Methods

Our approach at LAPDC is to look for subjects with some potential psychokinetic capabilities (PKers), thus able to produce rather regularly some small effects (the word "rather" is important here as the effects are not systematically produced). The difficulty when working with small effects is that a small effect implies a small signal, making it harder to detect and separate from the environmental noise. This is the well-known problem of the signal-to-noise ratio (SNR).

As explained in the ***Our Objective*** section above, the focus of this paper is on the analysis of aerodynamic forces. Other potential factors that can move a light mobile have been discarded, including magnetic and electrostatic forces, vibration, and radiation.

- Magnetic forces were eliminated by choosing a mobile made out of plastic.
- We performed tests with electrostatic forces (Figure M0e) but were unable to drive the mobile more than a half turn (the mobile swung between two positions and then stabilized itself).
- We obtained the same results with forces induced by mechanical vibrations. Vibrations from 10 Hz to 20,000 Hz were generated with a loudspeaker and measured using a smartphone accelerometer (Figure M0v). We never observed a movement larger than a half turn.
- For the effect of radiation, its impact is to be combined with the aerodynamic thermally induced effect analysis, radiation pressure being marginal.

For further details on the tests above you can consult LAPDC's website on the physical approach: https://sites.google.com/site/lapdctk1/LAPDC_Protocole_1a/hypotheses-classiques.

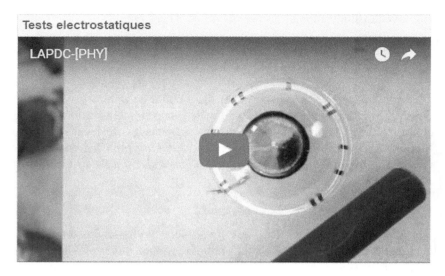

**Figure M0e. Electrostatic test with a generator.**

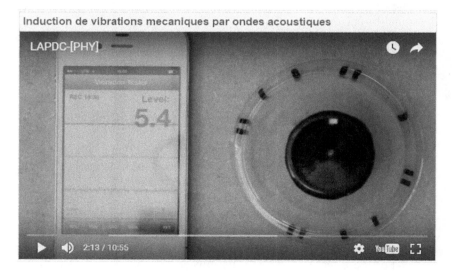

**Figure (M0v). Vibration test using an accelerometer.**

We developed some specific experiments to separate psychokinetic effects (if existent) from aerodynamic and thermally induced effects. The method does not require that the mobile be confined. This enlarges the effects we can observe (in particular, the observed speeds are greater in a non-confined environment). We attempt to model the details of the physics involved, to gain a deep understanding of the results and to be able to investigate their potential causes.

### Experiments Explained in This Paper

In this paper we will focus on one specific type of experiment using the same kind of spinning object. The aim is to compare the speeds of the air flows surrounding the object to the speed of the object itself. The complete approach for each experiment is detailed below.

Three different kinds of experiments have been compared:

- experiments producing mobile movement with an airstream (called A experiments),
- experiments in which the mobile spin is produced by a motor-drive (called M experiments),
- experiments which produced movement when a PKer put their hand close to the mobile (called PK experiments)

In this paper focused on the methodology, only A and M experiments will be presented.

The A experiments will show what kind of aerodynamic effect is able to spin the mobile. M experiments will show what happens to the air-flow speeds around the mobile when the mobile is spun by a small electric engine (and not aerodynamic effects). Besides serving as reference experiments against which the macro-PK test experiments (PK) could be compared, these A and M experiments were also a way to test and validate the measurement chain process.

To compare the experiments, the following ratio will be evaluated: mobile speed / mobile periphery air-flow speeds.

Indeed, as an energy transfer principle, if it is only aerodynamic effects that drive the mobile, then the air-flow speeds around the mobile have to be greater than the mobile speed, at least in some areas. So, if the value of the above ratio is smaller than 1, then the air-flow speeds on the mobile periphery are greater than the mobile speed, and we can infer that the air stream is able to drive the mobile. Therefore, aerodynamic forces are the probable cause for the mobile movement. If the value of the above ratio is larger than 1, the air-flow energy is insufficient to drive the mobile at this speed, and other explanations apart from aerodynamics forces must be found. Indeed, there is always some kind of energy waste in the transfer of energy between the air flows and the mobile, so a ratio equal or close to 1 is not sufficient.

**Figure M1. Hemispherical mobile device used for the experiments.**

### *Hemispherical Mobile Device (HM)*

The first step was to choose a standard mobile device. A plastic dome (with a hemispherical shape), easily found in retail shops, was chosen (Figure M1). The shape and the smooth surface give little grip for the air flow to take hold. This dome is placed on top of a needle inserted in a plastic support or cork. The weight of the dome is 2.4 g and its diameter is 85 mm.

### *Air-Flow Speeds Measured by PIV*

To analyze the impact of aerodynamic forces on the mobile, the first task is to be able to measure the air-flow speeds around the mobile. In this context, we work with low speeds (a few cm/s to 20 cm/s). Through collaboration with the Pprime laboratory of Poitiers University, the particle image velocimetry (PIV) technique was selected as the way to measure these speeds.

The principles of the PIV experiment we used are as follows (Figure M2):

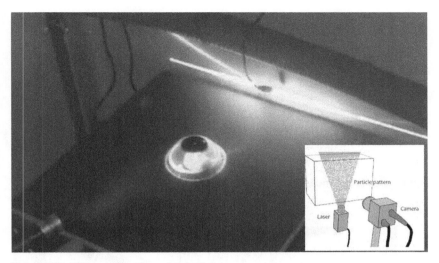

**Figure M2. PIV test bench and principle.**

— a laser beam illuminates a thin horizontal air slice just below the hemispheric mobile (HM) or at a specific height on the mobile,
— smoke is introduced (with a classic fog machine and some dispatcher pipes, sometimes helped by a small fan),
— the laser light diffuses on the smoke particles or/and smoke particle aggregates,
— a camera records the images (in our lab we use a 1080 x 720 definition sensor at a speed of 50 images/s),
— specific image processing software is used to deduce the speed of the smoke particles using their position change between two sequential images.

At LAPDC, we use PIV LAB, a MATLAB application developed by Thielicke and Stamhuis (2014). With this software we perform a pattern analysis using a fast Fourier transform (FFT) correlation algorithm, with multiple passes and deforming windows. Briefly, the software starts from the image pixels, compares two images, and tries to recognize identical patterns (particles aggregates or particles) in these two images (which may have rotated). Once some identical patterns have been identified, the software computes the pattern shift between the two images (Figure M3: $\Delta X$, $\Delta Y$). By using the time elapsed between the two images ($\Delta t$), the software is able to calculate the pattern speed and direction (U,V).

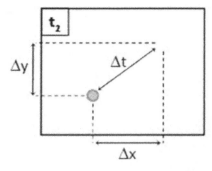

$$U(x,t) = \frac{\Delta x}{\Delta t}$$

$$V(y,t) = \frac{\Delta y}{\Delta t}$$

**Figure M3. Speed measurement principle.**

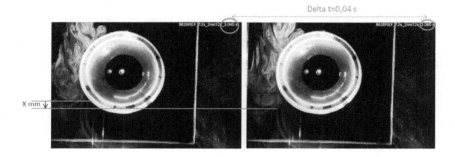

**Figure M4. Speed measurement principle used by PIV LAB.**

To illustrate the process, we present (Figure M4) two sequential images separated by 40 ms ($\Delta t = 0.04$ s). The smoke cloud front moved X mm during this time, so the speed of the point situated at the front of the cloud is X/0.04 = 25 * X mm/s. In this case there was a vertical shift only in the V direction.

Using the same process with all the identical patterns that could be detected, the software identifies the speed at many different points. It then calculates, by interpolation, a matrix representing the speed vectors field. Figure M5 gives a representation of the speed vectors field between these two images of the sequence. The narrow green and orange arrows represent the speed vectors at different points of the image. Orange arrows were obtained by interpolation when information was not sufficient for the system to compute the speed at certain points.

The software makes this process for image 1 and image 2, then image 2 and image 3, then 3 and 4, and so on. After that, it is possible to check some speeds at a specific time and a specific position or to ask for the mean speed vectors field on the total set of images, as shown in Figure M6.

**Figure M5.  PIV results: speed vectors field between two sequential images.**

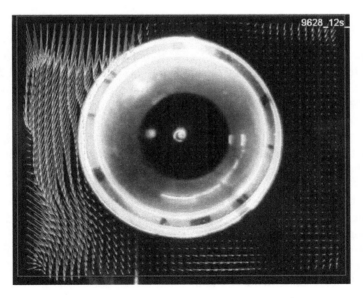

**Figure M6. Vector-field representation of the mean air-flow speeds around the mobile.**

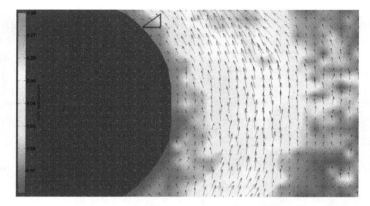

**Figure M7.  Color representation of the air-flow speeds around the mobile. Mobile is in red/brown. Other colors are associated with the range of air-flow speeds.**

Tools are available to study some parts of the air-flow movement in detail, or to show quickly, with color, the areas with a specific range of speeds (Figure M7).

### *Scan-Flow-Mobile Software*

The preceding analyses are interesting to check air-flow speed at a specific time, in a specific area around the mobile, or the mean air-flow speed during a time interval in a specific area. However, in order to analyze the air-flows speed evolution with time around the mobile, compared with the mobile-speed evolution, further tools were necessary. So we developed a set of procedures using MATLAB, which we named Scan-Flow-Mobile (Figure M8). They enable us to model the complete experience (movement of air and the mobile) as a numerical object and zoom in on some parts of the process (in space and time) with different granularity levels. With this tool we can compare different experiments and perform a fine analysis of the interaction between air flows and the mobile, and identify cause-and-effect relationships.

The first step is to extract from the previous calculation (done with PIV LAB) the tangential speed on concentric circles around the mobile. Each circle is represented with 100 points that are equidistant from each other. The distance between each circle is 1.5–2 mm. In parallel, a calculation of the curve representing the mobile speed evolution is performed using the software Tracker. The starting point is the same video used by PIV LAB to

# Scan flow mobile process

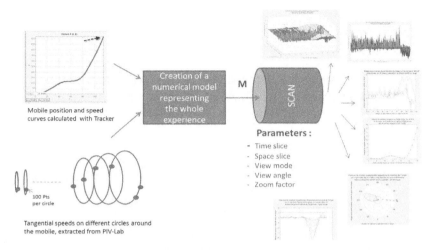

**Figure M8.  Scan-Flow-Mobile process.**

process the air-flows speeds. In the software Tracker, one specific marker on the mobile is identified every second, so the software can compute the corresponding angular position and angular speed curve.

Then, a numerical model is built to synchronize all of these data as a function of time. Extraction of one minute of data at a frequency of 50 Hz represents nine million points. Once this model is in place, it is possible to choose different ways of representing and extracting the data: speed surface representation with time, mean circle speed representation, different levels of granularity, linear or polar views, etc. A zoom can be performed on a specific part of the data in a linear or polar way.

### Experiments and Diagram Chosen for Results Presentation

These software procedures were used to perform a full analysis of the different experiments presented in the Results section. Each experiment has a report attached with the corresponding detailed results (internal reports at LAPDC that are identified by experiment number). We wanted to be sure that all factors capable of having an impact on the air flows and the mobile were taken into account. Sometimes, when looking at the results, we had to return to the original video in order to understand what had happened. (Example: discovering that the introduction of smoke had an impact on the air-flow speed in a specific way.)

As described before, we conducted different types of experiments A, M, and PK from 2014 to 2016:

— A experiments using air flows as the driver of the mobile movement,
— M experiments using a motor (small electric engine),
— PK experiments where a person, with some supposed specific skills, was able to set the HM mobile in motion by placing their hand close, without touching it. (These experiments are not presented here as we are focusing on the methodological part.)

We will present the results of these experiments in the Results section, and we will compare them with each other. To do that, one diagram was extracted from each individual experience, which represents the mean air-flow speed close to the mobile (1.5–2 mm) compared with the mobile speed. The mean air-flow speed is evaluated as follows:

— On each experiment, an "impacted zone" is identified (the angular zone on the circles where the flow is focused). In some cases, there is no impacted zone and the complete circle is taken into account. This is done to avoid any side effects on the mean speed calculation that could lower the real air-flow speed
— On each concentric circle a mean value is calculated on the "impacted zone" at each time step, usually 20 ms
— So, for a specific circle, we have the curve as a function of time of this mean value (blue curve).
— We chose the circle which is at a distance of 1.5–2 mm from the HM mobile periphery in order to have a good idea of the air-flow speed close to the mobile (typically this is circle number 25, 26, or 27 depending on the experiment setup).

The mobile speed is obtained as described before (integration in the numerical model of the results obtained with the software Tracker).

### *Looking for Clear Anomalies versus Significant Deviations from the Null Hypothesis on a Set of Calibrated Experiences*

For the use of this methodology to conduct a macro-PK test, as we defined in the Introduction (macro-PK or micro-PK), we do not need to use statistical tools to evaluate the rejection or not of the null hypothesis. In fact, we are looking for anomalous physical events inside the experiments. Contrary to a statistical approach, where experiments must be calibrated and repeated in the same design, using the same protocol, and in which all data segments have to be considered in order to avoid an observer bias, here we are performing experiments and checking (in each case) if we can find some part where an anomalous physical (AP) event clearly appears.

This macro-AP is detected in our PK experiment if the ratio mobile speed / mobile periphery air-flow speed is significantly larger than one over the whole experiment or for a part of it. Indeed, as the other physical factors are already discarded (see Introduction), and this ratio being greater than 1 also discards the aerodynamics factor, then the macro-AP is detected. There is no need for statistics because this event alone is an anomalous perturbation (provided that a calculation of potential errors has been correctly done—see Discussion).

In the case that only a part of the experiment[4] is selected to bring to light this ratio above 1, it is necessary to show that the ratio reversal was not induced by a preceding event (for example, by a launching of the mobile at high speed by other means). Thanks to the PIV techniques used, it is easy for the outside observer to do this check, as any impact on the air flow or the mobile are recorded and can be followed and checked at a rate of 50 frames per second.

## Results

### *A Experiments: HM Mobile Motion Driven by Generated Air Flows*

To evaluate the ways to set the HM (described previously) in motion, three different experiments using air-flow generation (with pumps and/or fans) were set up. It is sometimes complex to set a mobile in motion in a steady fashion with a generated air flow in an open environment. The first task was to be able to start the motion of the mobile and evaluate what air-flow speed was required to achieve a mobile speed between 10 and 40 deg/s (typical speed reached with PKers). Therefore, the first two experiments used a mechanism with a pump aspirating a mix of smoke and air. The output of the pump was pushed inside a hose (small pipe such as the kind used in an aquarium), with one or several exits, bringing the air flow to the PIV bench (Figure R0). The third experiment was intended to study the minimum air-flow speed required to keep the mobile running, in a PK-like environment; it used a fan with separate input for the smoke introduction.

In the following sections, many images are used to describe the experiments and their results. These images mainly give a view from above, so the HM mobile appears as a disk.

**Experiment A.1: Orthogonal air flow**. The first experiment (#9628), performed in September 2015 on David Jamet's PIV bench, used the pump mechanism described above with a hose split for two exits, each one with

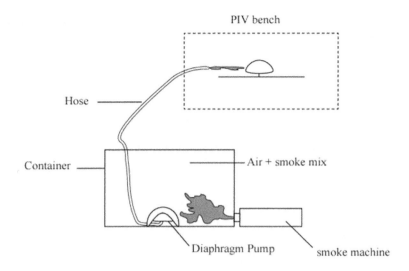

**Figure R0. Mechanism for air-flow generation in A-1 and A-2 experiments.**

a small tap. The first exit represented the main air flow, which we can see on the left of the HM, coming from the top of the figure (vertical flow on Figure R1). The second exit was horizontal and on the left, orthogonal to the preceding one (with the flow direction from left to right). Mobile motion has been produced with these two flows combined. The experiment duration was 120 s. After air-flow speed calculations using PIV LAB, the vector-field representation of the mean airflow speed around the mobile during these 120s is shown in Figure R2.

After Scan-Flow-Mobile processing, the evolution of the mean air-flow speed at 1.5 mm from the mobile (blue curve) compared to the mobile speed (red curve) during the 120 s duration is presented in Figure R3. The mobile began to move at about 65 s and reached a speed of 9.3 deg/s. To generate this motion, the mean tangential air-flow speed (converted to rotational speed) oscillated in the range 20–70 deg/s (with an approximate global mean of 40 deg/s).

Remarks: The global pump mechanism (smoke generation, pump, hose, and hose split) and the turbulence at the hose exits generated some speed oscillations that we can see on the graph (oscillation period in the range of 12–18 s). This experiment was insightful but not very efficient in the generation of motion. The mobile first oscillated around its axis without starting a circular motion. Then, after about 65 s, it began to move with a global flow which was a little stronger and better synchronized. In reality,

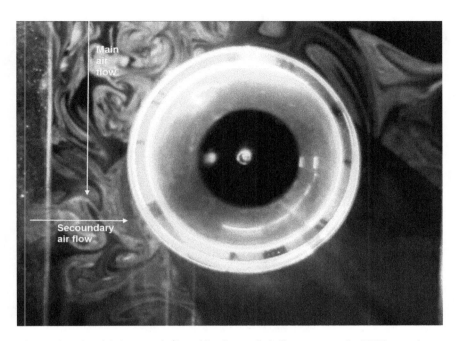

**Figure R1. Combining vertical and horizontal air flows to set the HM in motion.**

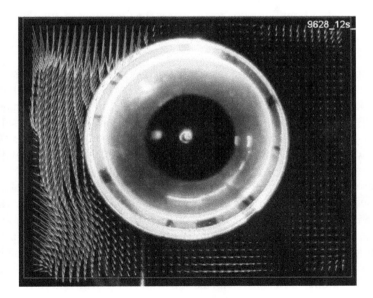

**Figure R2. Vector-field representation of the mean air flow speed around the mobile.**

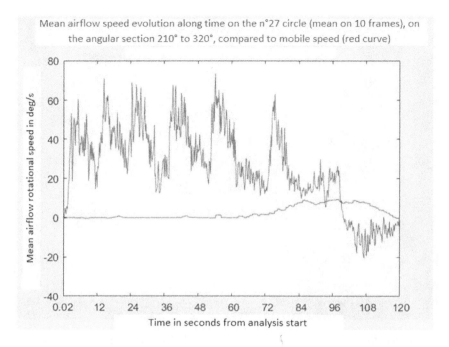

Figure R3. **Mean air flow speed close to the mobile compared with the mobile speed (red). The horizontal axis is the time elapsed and the vertical is the rotational speed in deg/s.**

the experience ended at 90 s, as the 10 deg/s mobile speed was reached. The portion of the graph after 90 s is not very meaningful: The first tap was closed allowing flow through the second exit, close to horizontal but slightly upward, which means that the air flow direction was reversed and the air-flow speed became negative. This explains the blue curve after 100 s.

**Experiment A.2: Focused tangential air flow.** To improve the efficiency of motion generation, a second experiment (#1251) was performed in February 2016 on the Davis Jamet PIV bench; only one focused tangential airflow was used (Figure R4). The recording duration was 84 s. Here again the focus was on starting the motion of the mobile. Unlike in experiment A.1, the hose is not split and is close to the mobile (light green part on the left of the image) which gives a more focused (air jet) air flow, which is very efficient at driving the mobile.

After the air-flow speed calculation with PIV LAB, Figure R5 shows the vector-field representation of the mean airflow speed around the mobile, during the 84 s of the experience.

**Figure R4. Focused tangential airflow to set the HM in motion.**

After Scan-Flow-Mobile processing, Figure R6 shows the mean air-flow speed evolution at a distance of 1.5 mm from the mobile (blue curve) compared with the mobile speed (red curve) over 84 s. The mobile started to move very quickly and reached a speed of 39.5 deg/s. To launch this motion, the mean tangential air-flow speed (converted to rotational speed) oscillated in the range 50–120 deg/s, with an approximate global mean of 75 deg/s during the launch period.

Remarks: The generation of motion was very efficient. A speed close to 40 deg/s was reached in just 50 s. As soon as the air flow had been stopped (56 s), the mobile took 25 s to return to being still (due to inertia and the action of air and needle friction).

**Experiment A.3: Air flow generated by an axial fan with an artificial hand as the obstacle.** The third experiment (#4327) took place on David Jamet's PIV bench in February 2017 (Dullin & Jamet 2017). One air flow was generated by an axial fan (not very focused), and a wood artificial hand was used to simulate the obstacle created by a hand in the air flow (Figure R7).

The idea behind this effort is to perform a type A experiment (mobile motion initiated by generated air flows) but with some PK experiment characteristics: a not very focused flow (axial fan) with a hand as an obstacle (a wood artificial hand). The mobile movement is induced, as in the Warcollier (1908) explanation, by the dissymmetry of forces around the HM; only one side is driven by the air flows while the other side is "protected" from the air

**Figure R5. Vector-field representation of the mean air flow speed around the mobile.**

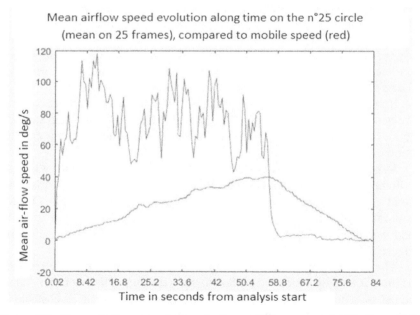

**Figure R6. Mean air-flow speed close to the mobile compared with the mobile speed (red). The horizontal axis is the elapsed time, and the vertical axis is the rotational speed in deg/s.**

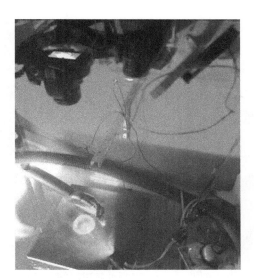

**Figure R7. Experimental setup with an artificial hand.**

flows by the hand (Figure R8). The difference from the Warcollier experiment is that the air flows here are generated by an axial fan.

Working on the air-flow speeds, we tried to find the speed limit above which the mobile starts moving and maintains a steady rotation. An 18-s period of time when steady movement of the mobile was obtained was chosen. After air-flow speed calculations using PIV LAB and Scan-Flow-Mobile processing, Figure R10 shows the mean air-flow speed evolution 1.85 mm (blue curve) and 3.70 mm (orange curve) from the mobile compared with the mobile speed (red curve). The mobile speed was approximatively 4 deg/s. The mean tangential air-flow speed was 1.85 mm from the mobile oscillated from negative values (reverse flow during a few seconds around $t = 11$ s) to 100 deg/s with a global mean around 22 deg/s.

Remarks: As it is not generated using same mechanism as the other two A experiments, the same air-flow speed oscillations do not occur. The air flow here is much more evenly distributed (even if we can find a periodicity of about 2 s in the peaks). Several factors could explain this: the use of a fan, placed on the left of the image, rather than a focused air flow (so farther from the mobile) and the obstacles in the flow. The elementary peaks (every 2 s) in the air-flow speed do not impact the mobile speed because of its inertia. The air flow had a global impact on maintaining the mobile speed at an average value of 4 deg/s, which was the lowest speed that can be obtained with a steady movement. So, as expected, the efficiency was low: We achieved a speed of 4 deg/s for a mean speed of about 22 deg/s with some high speeds around 80–100 deg/s.

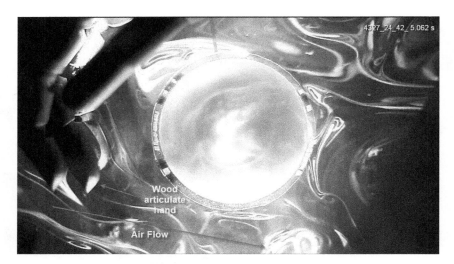

**Figure R8. Experimental setup with an artificial hand. Direction of the air flow, an artificial hand as an obstacle to a part of the airstream, leaving a tangential air flow on the other side of the mobile.**

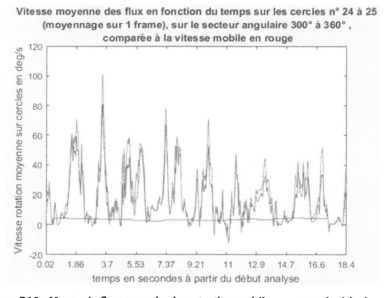

**Figure R10. Mean air-flow speeds close to the mobile compared with the mobile speed (red). The horizontal axis is the elapsed time, the vertical axis is the rotational speed in deg/s. Blue curve: air flows 1.85 mm from the mobile; orange curve: air flows 3.70 mm from the mobile.**

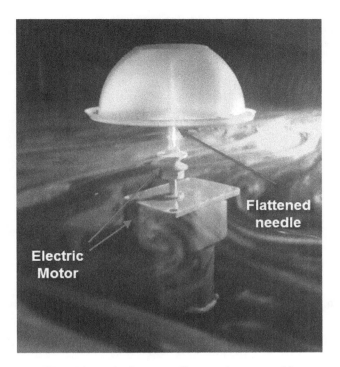

**Figure R11. HM mobile attached to a needle spun by motor drive.**

### M Experiments: HM Motion Driven by a Motor

In these two experiments, the HM mobile is driven into motion by a motor. In fact, the HM is placed on a needle (flattened a little on the top) with some "patafix" stuck on the mobile. This needle is then spun by a motor, so the spin drives the mobile, too (Figure R11).

**Experiment M.1.** This experiment (#2331) was performed in August 2016 on David Jamet's PIV bench. After air-flow speed calculations using PIV LAB and Scan-Flow-Mobile processing, Figure R12 shows the evolution of the mean air-flow speeds at the mobile periphery (and also at some farther distances from the mobile), compared with the mobile speed (red curve), over 52 s. The mobile driven by the spun needle achieved a speed of 25 deg/s and then 50 deg/s. The mean air-flow speed on circle 25, which is 1.8 mm from the mobile, was approximatively 5 deg/s. After the fast starting of the mobile motion, the air flows accelerated with a delay of between 0 and 7 s. It appears also that the air flows closer to the mobile were faster than those farther away. We can therefore conclude that the air flows were lightly driven by the mobile.

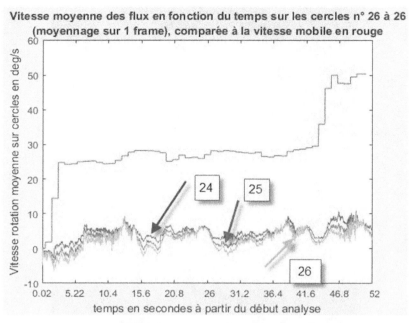

**Figure R12. Mean air-flow speeds close to the mobile compared with the mobile speed (red). The horizontal axis is the elapsed time since the beginning; the vertical axis is the rotational speed in deg/s. Curve 24: mobile periphery; curve 25: 1.8 mm from the periphery; curve 26: 3.6 mm from periphery.**

**Experiment M.2: Motor-driven mobile with a large acceleration on the second part**. This second experiment (#3039) was performed in November 2016 on David Jamet's PIV bench. It was done to confirm the first experiment and also to further study the air-flow behavior in the case of strong acceleration of the mobile. After air-flow speed calculations using PIV LAB and Scan-Flow-Mobile processing, the evolution of the mean air-flow speeds 1.8 mm (curve 24) and 10.6 mm (curve 30) from the mobile compared with the mobile speed (red curve) over 63 s is shown in Figure R13. The mobile, driven by the spun needle, reached a speed of 17 deg/s, then stopped at $t = 42$ s. Then, with a very strong acceleration, it quickly spun up to 70 deg/s. Accordingly, the mean air-flow was initially around 5 deg/s and it then increased to 22 deg/s. Air entrainment by the mobile is clearly visible here: The air-flow speed closer to the mobile periphery (circle 24) is higher than farther out (circle 30). When mobile acceleration occurred, air-flow speeds became greater also.

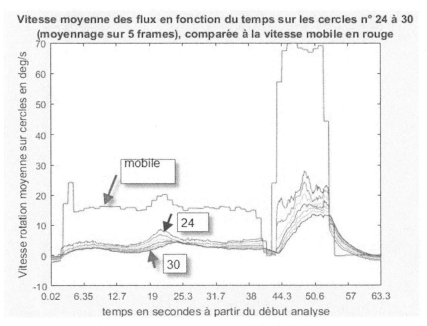

**Figure R13. Mean air-flow speeds at different distances from the mobile compared with the mobile speed (red). The horizontal axis is the time elapsed since the beginning, the vertical axis is the rotational speed in deg/s. Curve 24: at 1.8 mm from the mobile periphery; curve 30: at 10.6 mm from periphery; other curves: between.**

Figure R14 shows the air-flow speed vector field, calculated between two images, during the period of strong acceleration. The mobile is driven in a clockwise direction and drives the air-flows (green arrows). However, it is important to notice that the mean air-flow speed remained much lower than the mobile speed, as in the M1 experiment. In the comparison part, this experiment will be divided into two parts: "low speed" from 0 to 42 s and "high speed" from 42 to 63 s.

### Comparison between Results of A and of M Experiments

To compare the preceding experiments, we chose in the Methods section to evaluate the ratio: mobile speed / mobile periphery air-flow speed. We took as periphery air-flow speed the one situated at least 1.5 mm (A1 and A2 experiments) from the mobile periphery, to avoid potential calculation errors (see Discussion: DPIV algorithm, where DPIV is digital PIV). Six results are presented in the table below comparing mean mobile periphery

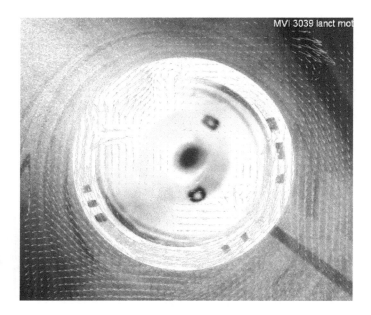

**Figure R14. PIV results: speed vectors field between two sequential images
separated by 48.735 s.**

air-flow speed (named "Air-flow speed"), mean mobile speed (named "Mobile speed"), and the ratio between the two (Table 1).

The 6 results are placed on the diagram in Figure R27. The different experiments are placed on the horizontal axis in the order used to present them above. They are grouped by category (A, M). The vertical axis represents the ratio value.

The diagram shows a clear separation between air-flow experiments (A), for which the ratio is lower than 1 (max 0.53), and the motor-driven category (M), for which the ratio is above 1 (min 2.92). These consistent results across different experiments contribute to the validation of the measurement protocol.

## Conclusion

The methodology proposed here provides a way to identify if the aerodynamic effects can be or not be the only cause of the spinning of a light object in a non-confined environment. In the experiments presented, and the Figure R27 outline of the M category of experiments, aerodynamic effects could not explain the mobile spinning (which is correct as the movement is due to the torque of the electric motor on the needle).

**TABLE 1**
**Results Comparison between the Two Categories of Experiment**

|  | | | | | | |
|---|---|---|---|---|---|---|
|  | 40 | 75 | 22 | 5 | 5 | 24 |
|  | 9.3 | 39.5 | 4 | 28 | 17 | 70 |
| Ratio | 0.23 | 0.53 | 0.18 | 5.60 | 3.40 | 2.92 |

The methods developed provide many tools with which to study in more detail what is really happening to the air flows and the spinning object during the experience. Also, the techniques proposed can easily detect any trickery involving aerodynamic forces used to move the mobile (such as mouth air blowing or hand movements).

This approach could be a good candidate to evaluate macro-PK tests in a non-confined environment. As this set of experiments constitutes a pilot study (the first time this kind of technology is used on this topic), the technology and the protocols used need to be confirmed and improved by other experiments in order to confirm the results obtained. We are looking to improve the total measurement process documentation, so that other laboratories can test it in their experiments.

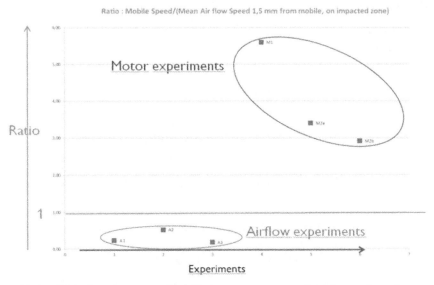

**Figure R27.  Results comparison between two categories of experiment.**

## Discussion

### *Plan Approach versus 3D Approach*

One could argue that the slice chosen to make the laser plan might not be representative of the air flows around the mobile on other slices. To answer this point, we can say that:

— We are not in a turbulent mode (speeds are low), so air flows are homogeneous around the mobile. There is no shear fracture in the flow vertically. If this was the case, we could detect it with a thorough analysis of the PIV results in the plan (many patterns would disappear from one image to the next).
— A experiments are processed in the same way as those performed using PKers, and we need to have the air-flow driving force in the same plan as the laser to be able to analyze it.

Furthermore, in our current research, we also have conducted some experiments with vertical PIV (to investigate the vertical air flows). These experiments have given no indication of flaws in the method described in this paper.

### *Tolerance on Ratio Evaluation*

To compute the possible error in the ratio evaluation that we present in the Results section, we must look at possible sources of error:

The uncertainty parameters associated with the air-flow speed calculations such as:

- the difference between fluid and particle velocity,
- image acquisition error and frequency acquisition tolerance,
- calibration,
- the DPIV algorithm (image treatment),
- impacted zone and circle mean speed determination,
- mean choice to be used in the ratio for the global experience.

The uncertainty parameters associated with the HM mobile speed calculation such as:

- mobile image acquisition error (same as for the air flow),
- marker angular position measured by the software Tracker,
- mean choice to be used in the ratio for the global experience.

We will now look at some of these parameters.

### Difference between Fluid and Particle Velocity

In a continuously accelerating element of fluid, the velocity lag $U_s$, between particle velocity $U_p$, and fluid velocity U, can be written using Stokes' drag law, as in Figure D1 (Raffel et al. 2007), where $d_p$ is the particle diameter, $\rho_p$ is the particle density, $\rho$ is the fluid density, $\mu$ is the fluid dynamic viscosity, and $a$ is acceleration.

$$U_s = U_p - U = d_p^2 \frac{\rho_p - \rho}{18\mu} a$$

**Figure D1. Stokes' drag law.**

The size of the elementary smoke patterns followed by PIV LAB in our experiment are estimated to be in the range of 1–8 pixels. So, in the worst-case scenario, we have 4 pixels as $d_p$. In most of our calibrations, 1 pixel corresponds to 0.18 mm, so $d_p$ = 0.72 mm. Smoke particles are made of glycol which has a density of 1.110 $kg/m^3$. Using the parameters for air at 20 °C, we obtain $U_s$ = 0.4a. The mean air-flow acceleration in the A experiment is close to zero, sometimes even negative (the only changes in speed are due to oscillations around the same value because of the experiment setup) (Dullin 2015–2017). Therefore, this parameter has no effect.

At the beginning of the M1 experiment, the air flow increases from 0 to 8 deg/s in 7 seconds, so the acceleration is 1.14 (deg/s)/s. That gives a speed error of 0.4 * 1.14 = 0.46 deg/s (the speed of the air increases by 0.46 deg/s).

In the second part of the M2 experiment, the speed increases from 0 to 20 deg/s in one second (blue curve). So, $a$ = 20 (deg/s)/s implies $U_s$ = 0.4 * 20 = 8 deg/s, and the air-flow speed could be 28 deg/s instead of 20 deg/s. We can easily correlate this with the next part of the curve, which effectively peaks at 28 deg/s, as the smoke particles are now moving at almost the same speed as the air because the acceleration has returned to zero.

In the T experiment, only the T.1 part shows acceleration of the air flow (small), with the mean speed changing from 2.5 to 5 deg/s between 50 s and 75 s; this means that acceleration is 0.1 (deg/s)/s. So, the speed error is 0.4 * 0.1 = 0.04 deg/s, which is marginal compared with the other error factors.

In conclusion, this factor has no real impact on the results seen above. It increases the ratio for the second part of the M.2 experiment from 0.34 to 0.37 because of the very large acceleration of the motor drive.

### Image Acquisition Error and Frequency Acquisition Tolerance

A continuous wave laser is used in the experiments, alongside a good quality camera (CANON 600 D with a resolution of 1080 * 720 pixels. All experiments capture images at a rate of 50 image/s. Error at this level is marginal compared with the other error sources.

### Calibration

The same type of mobile is used for all experiments. Its diameter (85 mm) is used in the PIV LAB calibration (Figure D2). So, the uncertainty in its value is associated with the drawing used to define the diameter on the image. We can estimate this at 2 pixels on each side, giving 0.18 * 4 = 0. 72 mm or 0.72/85 = 0.8%.

### DPIV Algorithm (Image Treatment)

This is an extract from Thielicke's thesis (Thielicke 2014):

> In DPIV, the particle displacement is calculated for groups of particles by evaluating the cross-correlation of many small subimages (interrogation areas). The correlation yields the most probable displacement for a group of particles traveling in a straight line between image A and image B. Under optimal conditions, the bias error of the window deformation DPIV algorithms used is smaller than 0.005 pixels and the random error is below 0.02 pixels. When the average displacement in a DPIV study is around 6 pixels, the displacement error goes below 0.42%.
>
> In our case (small speeds), the average speed in an A experiment is around 40 mm/s (75 mm/s maximum in A2), which gives an average displacement of 4 pixels (8 pixels maximum) for a 50-Hz image capture frequency and a pixel size of 0.18 mm. With M and T experiments, the average speed is more like 10 mm/s, which gives an average displacement of 1 pixel for the same image capture rate. So, we can assume an error of 1% for this part of the process.
>
> It is important to note that very close to the front of the object (closer than 1 mm), DPIV gives a decreased air-flow speed if the mobile is slower and an increased air-flow speed if the mobile is faster. However:
> The speed values are taken at least 1.5 mm from the mobile.
> If some error were due to this factor, our results are reinforced because the air-flow speed for A experiments should be increased (air flow faster than the mobile) and the air-flow speed for T experiments should be decreased (air flow slower than the mobile).

Further studies are required to evaluate the impact of the window interrogation choice (Thielicke & Stamhuis 2014) on the high air-flow speed treatment (for A experiments). In fact, the speed values in A experiments could be higher than our current evaluation (which would further reinforce our results). Work has already been done on the A2 experiment with the higher air-flow speed input.

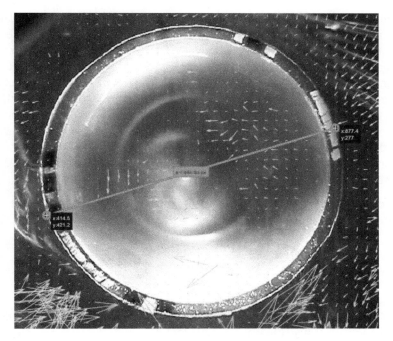

**Figure D2. Calibration in PIV LAB using HM mobile diameter (85 mm).**

### *Impacted Zone and Circle Mean Speed Determination*

There is uncertainty in the exact impacted zone chosen when computing the mean speed on the circle (start angle, end angle) at each time step; the estimated error is ±2%.

When we evaluate the mean speed to be taken into account for the air flow from the preceding curves, even if we also use a software calculation to confirm it, there is some uncertainty depending on the part of the curve taken into account. Some other operator could have chosen a slightly different value. We estimate the error to be ±3%.

So we have a combined error risk of ±5% which gives a 10% potential error in the air-flow speed.

### *Marker Angular Position Measured in Tracker and Mobile Speed Evaluation*

Using the software Tracker, we pinpoint the marker position at each second of the experiment. The software deduces the cumulative mobile angular position and its angular speed at each time stamp. When we locate the marker position, there could be a ±1-degree error in the angular position, although this is not cumulative, as we redefine the exact position of the marker each second. So, between two points there could be a 2-degree error in the measurement which impacts the speed calculation between these points. However, at the next point this error will be corrected in the other direction because we always process a new measurement for each step. We conclude that the main error happens when we evaluate the mean of the global curve in order to calculate the ratio. We have estimated this risk to be the same as for the air flow (±3%), implying a global risk of 6%.

### Conclusion: Error Estimation on the Ratio

To calculate the uncertainty on the mean air-flow speed, we add the following individual errors:

> — 0.8% for the calibration,
> — 1% for the DPIV treatment,
> — 10% for the mean calculation,

which gives a total error of 11.8%. For the mean mobile speed, we calculated an error of 6%. Finally, we have an uncertainty of 11.8 + 6 = 17.8% (±8.9%) in the value of the ratio.

### Consequences of the Ratio Diagram

The highest ratio in the A experiment (A2) is 0.53; 8.9% above this point gives 0.58. The lowest ratio in the M experiment has a value of 2.92; 8.9% below this point gives 2.68. Therefore, we have a ratio of 5 between the two closest points of the A and M categories. The uncertainty error in the ratio cannot explain the differences between these two categories (driven by air flows and motor-driven).

### Acknowledgments

We wish to thank all the people and organizations who, directly or indirectly, gave support to the LAPDC project and contributed to the development of this method and the acquisition of results. More specifically, we would like to thank the CNRS–Pprime laboratory, in particular the AFVL[5] president Laurent David, the entire LAPDC PKers volunteer team, private donors, and the IMI[6] foundation. We also wish to thank Renaud Evrard for his reading of the article and his editorial comments for the improvement of this paper.

## Notes

[1] TK: At the LAPDC we use TK for telekinesis. For this paper, we will use PK for psychokinesis. This point is discussed at the end of the Introduction.

[2] As an example, here is a search on Youtube with the *psychokinesis* keyword: https://www.youtube.com/results?search_query=psychokinesis

[3] As an example, here is a search for psi wheels: http://www.internationalskeptics.com/forums/showthread.php?t=117190&highlight=psi+wheel

[4] Sections of the experiment are also chosen according to measurement conditions (for example, a video showing good lighting and good smoke-particle seeding).

[5] AFVL: Association Francophone de Vélocimétrie Laser (Laser Velocimetry French-speaking Association).

[6] IMI: Institut Métapsychique International.

## References Cited

Alvarado, C. (2006). Human radiations: Concepts of force in mesmerism, spiritualism, and psychical research. *Journal of the Society for Psychical Research, 70.3:*138–162.

Archat, P. (1908). Recherche expérimentale de l'action motrice sans contact. *Annales des Sciences Psychiques (section Actes de la Société Universelle d'Études Psychiques), 17–18*(16 June to 1 July):198–201.

Auerbach, L. (1996). *Mind over Matter.* Kensington.

Dullin, E. (2015–2017). Experiment setting and report analysis (9628, 1251, 4327, 2331, 3039, 81L-Part 2, 7076 start, 7076 launched, 2756). *LAPDC Internal Reports.*

Dullin, E., & Jamet, D. (2017). Du moteur à fluide à la PK sur le net (From the bioforce motor to the PK on the Internet). *Métapsychique, 1:*76–85.

Egely Wheel (2017). http://egelywheel.net/

Evrard, R. (2016). *Enquête sur 150 ans de parapsychology.* Editions Trajectoire.

Giuditta, A. (2014). On the human bioenergy field. *WISE Journal, 3*(3).

Hofmann, A. (1919/1992). *Das Rätsel der Handstrahlen: Eine Experimental Studie.* Leipzig: Oswald Mutze.

Hansen, J. N., & Lieberman, J. A. (2013). Use of a torsion pendulum balance to detect and characterize what may be a human bioenergy field. *Journal of Scientific Exploration, 27*(2):199–219.

Jahn, R. G., & Dunne, B. J. (2011). *Consciousness and the Source of Reality—The PEAR Odyssey.* Princeton, NJ: ICRL Press.

Krmessky, J. (1975). Psychokinesis Research in Czech–Soviet and Czech parapsychology research. U.S. Defense Intelligence Agency, declassified, p. 47.

Kruth, J. G. (2016). Experimental PK Studies with Exceptional Participants—Rhine Research Center. *Parapsychological Association Convention* poster.

Martin, C. (1926). Essais expérimentaux sur les effets moteurs du supposé fluide magnétique humain. *Revue Métapsychique, 2:*125.

May, E. C., Utts, J. M., & Spottiswoode, S. J. P. (1995). Decision Augmentation Theory: Toward a model of anomalous phenomena. *The Journal of Parapsychology, 59*(September):195–220.

Perelman, Y. (1913). *Physics for Entertainment* [in Russian]. [Reference is made to a communication to the Moscow Medical Society from N. P. Nechayev in 1876, entitled "The Gyration of Light Bodies Caused by the Heat of the Hand."]

Raffel, M., Willert, C., Wereley, S., & Kompenhans, J. (2007). *Particle Image Velocimetry (2nd Edition)*. Springer Publishing,

Rudkin, J. (2001). The Egely Wheel—A Telekinesis detector? *Society for Psychical Research Paranormal Review, 19*(July).

Stine, G. H. (1985). *On the Frontiers of Science—Strange Machines You Can Build*. Collier Macmillan.

Thielicke, W. (2014). The Flapping Flight of Birds—Analysis and Application. Ph.D. Thesis, Rijksuniversiteit Groningen. http://irs.ub.rug.nl/ppn/382783069

Thielicke, W., & Stamhuis, E. J. (2014). PIV LAB: Time-Resolved Digital Particle Image Velocimetry Tool for MATLAB. https://figshare.com/articles/PIVlab_version_1_35/1092508

Tromelin, Comte de, G. (1907). *Les Mystères de l'Univers*—Collection Bon. Paris: Bibliothèque Universelle Beaudelot.

Tromelin, Comte de, G. (1912). Le fluide vital ou force biologue chez l'humain normal. *Le Monde Psychique-Juillet 1912 de l'institut de Recherches Psychiques de France*:432–439.

Varvoglis, M., & Bancel, P. (2015). Micro-Psychokinesis. In *Parapsychology: A Handbook for the 21st Century,* Chapter 20, edited by Etzel Cardeña, John Palmer, and David Marcusson-Clavertz, Jefferson, NC: McFarland.

Warcollier, M. (1908). Le "Moteur à fluide" du comte de Tromelin. *Annales des Sciences Psychiques, 1908*:201–203.

Watkins, G. (2015). Macro-Psychokinesis. In *Parapsychology: A Handbook for the 21st Century,* Chapter 6, edited by Etzel Cardeña, John Palmer, and David Marcusson-Clavertz, Jefferson, NC: McFarland.

*Journal of Scientific Exploration*, Vol. 32, No. 3, pp. 555–578, 2018          0892-3310/18

*RESEARCH ARTICLE*

# Prediction of Mental Illness Using Indian Astrology: Cross-Sectional Findings from a Prospective Study

RAJESHKRISHNA PANAMBUR BHANDARY

rajesh.kbp@manipal.edu

PODILA SATYA VENKATA NARASIMHA SHARMA

HEMA THAROOR

Submitted July 7, 2017; Accepted August 7, 2018; Published September 30, 2018

DOI: https://doi.org/10.31275/2018/1260
Copyright: Creative Commons CC-BY-NC-ND

**Abstract**—Studies involving astrology and psychiatry have mostly found conflicting results, with astrology being criticized as unscientific and also lacking an objective assessment method for being scientifically tested. We tested the predictive ability of astrology using the Indian system in identifying mental illness on 150 subjects (75 having mental illness and 75 without). Four astrologers blind to the subjects interpreted the computer-generated birth chart data derived from subject's gender, and date, time, and place of birth. Predictions were matched with the clinical details at first assessment. Kappa coefficients suggested a moderate agreement in the astrological prediction of lifetime mental illness ($k = 0.560$, $p = .001$) and a substantial agreement in predicting current state of mental illness ($k = 0.626$, $p = .001$), but with good inter-astrologer agreement only for lifetime mental illness. Viewed as a diagnostic test, astrology showed a good sensitivity and specificity for identifying mental illness of more than 75% for lifetime mental illness and more than 80% for the current mental illness. However, the study showed a poor match in predicting the symptom cluster and time of onset of symptoms. Overall, the evidence seemed to point toward Indian astrology (*Vārāhamihira* system) as practiced in the study modestly predicting the presence of mental illness. Caveats included differences in diagnostic concepts in the allopathic and astrological systems and the lack of a predefined level of astrological analysis which resulted in a poor match for symptom cluster and timing of illness. The findings in this study hence are at best tentative and need more extensive enquiry.

*Keywords:* Indian astrology—mental illness—prediction

## Introduction

Astrology has been considered a science in India since the time of the *Vedas* (ancient Indian texts). For various reasons, astrological sciences have been reduced to mere assumptions and chance predictions in the modern era. The basis for astrological prediction often appears illogical, unscientific, and indemonstrable in today's world. However, astrology still persists because it provides a coherent and comprehensive system of thought and gives an explanation for why each human being is different from all others. And astrology provides an explanatory model for the occurrence of illness (Hare 1977). Astrology forms one of the predominant treatment methods in India during mental illness. A study done in rural India by Kapur (1975) found that 59% of those with mental illness consulted traditional healers and practitioners of indigenous medicines. Out of these, astrology forms a major portion where the chanting of mantras and performing of certain rituals and prayers are a means of therapy. In ancient India, astrology was a good source for health promotion and a guide for preventive, curative, and other aspects of treatment which are mentioned in Ayurvedic (Traditional Indian Medicine system) literature (Sharma, Prasad, & Narayana 2007, Sharma, Subhatka, & Narayana 2007).

Almost all variants of astrology implicate the ill effects of the moon on mental illness. Evidence supports the impact of the lunar cycle on the physiology of animals and humans, in particular fertility, menstruation, and birth rate (Zimecki 2006). Studies involving epilepsy, which is seen as a mental illness in astrology, have found a clustering of seizures during the full moon (Polychronopoulos et al. 2006, Baxendale & Fisher 2008) However, one study by Benbadis et al. (2004) found only pseudoseizures to have such an association. With regard to lunar cycles and violence, Owen et al. (1998) found a positive correlation among inpatients, and another study by Hicks-Caskey and Potter (1991) found more misbehavior in developmentally delayed institutionalized women on days of the full moon. Behavioral changes were greater in those with psychotic illnesses as compared with other mental illness during the full moon (Barr 2000). However, Snelson (2004) criticizes these inferences in a study that found a correlation between higher belief scores and the attribution of behavioral changes in the lunar cycle.

Looking at birth charts using the Western astrology system, the Carlson Experiment did not find a match between people's acknowledgment of personality profile by psychological assessment and astrologers' predictions (Carlson 1985). A more recent study by Wyman & Vyse (2008) with Neo Five versus computer-generated astrological profile (as per Western astrological principles) found more of a chance association. On the other hand, a study by Abdel-Khalek and Lester (2006) who looked at sun signs (Western astrology) found higher anxiety scores in Scorpio, Aquarius, Gemini, Pisces, and Capricorn, and subjects with lowest scores in Libra, Taurus, Leo, and Cancer signs. Natal chart interpretations (analyzing zodiacal signs, interactions among planetary qualities as per Western astrology, and occurrence of full and new moon dates, on the dates of birth) have also shown some support for key astrological postulates with sun signs with introversion, and that Mars, a symbol of aggression in outgoing signs, is more often found in schizophrenics than normal (Ohaeri 1997). Ertel (2009) in his critical appraisal of Carlson's study had found many drawbacks regarding the methodology as well as analysis, including the tool used (California Personality Inventory), the analysis done on a piecemeal basis, and small effect sizes for the findings. Also, the author noted that the design of the study was unfair as it used a three-choice format over a two-choice format for discrimination.

The continuing problem with astrological predictions is that people may tend to associate themselves with profiles that are better than they are or that make them feel better. Although the practice of astrology is increasingly seen to be absurd, many believe that this is only because most of its practitioners are incompetent or fraudulent. This might still be the truth.

The center of this study has the same catchment area as in the study by Kapur (1975), where health-seeking is through astrologers for mental illness. Hence, it is of the utmost importance to understand this cultural influence while dealing with illness. While studies are restricted to the effect of the moon (lunar cycle) on suicide (Sharma & Thakur 1980) and emergency admissions (Zargar et al. 2004), no studies seem to have been done in India on how astrology as a whole concurs with the manifestation of a psychiatric illness. Hence, it was felt that an investigation into the detection of mental illness and the prediction of outcomes between the allopathic and astrological systems would be beneficial in helping us to understand the two perspectives better. Consequently, the present investigation was planned which also attempts to make the assessment a more objective one, using comprehensive knowledge of astrology and ensuring adequate blinding of subjects from astrologers and vice versa.

## Brief Introduction to Indian Astrology

Mythological claims are that Indian (Vedic/Hindu) astrology originated from Brahma (the creator of the world as per Hinduism) and was brought to earth by the sages. 18 such works of the sages were then compiled by Vārāhamihira in the Fifth Century AD to form the *Brihatjataka* (Great

Treatise on Birth History/charts), which forms the major source of reference for interpretation in Indian astrology.

The foundations for predictions about an individual lie in the plotting of the birth chart. The zodiac is an imaginary belt in the sky in which the planets travel. The zodiac is divided into 12 equal parts of 30° each which form the *Rashi* (or the zodiac sign). According to this system, there are 9 planets that have influence on the earth (and are identifiable from the earth). In addition to the 5 planets up to Saturn, the Sun and Moon are also taken as planets. In addition, there are two imaginary nodal points (at 180° from each other) called *Rahu* and *Ketu,* which are referred to as the shadow planets. The Zodiac is also divided into 27 constellations of 13°20′ each based on the position of the star, and each constellation is further subdivided into 4 quarters. The planets are plotted in reference to their position in the zodiac with reference to the position of the birth star as viewed from earth (hence this system is called sidereal). The time of birth determines the zodiac sign for the individual which can change within a day based on the position of the Sun. Depending on the need, the birth chart can have accessory charts that plot the planets into smaller divisions within the Zodiac, in order to reach a higher accuracy. Planets are assigned strengths based on the constellation they are in, the relative position from the zodiac sign, as well as based on the position of other planets in the Zodiac. While this forms the static chart at birth, the dynamic component is analyzed by the operational periods of the planets (which varies from 6 years to 20 years) across a timeframe of 120 years. Each operational period can be further divided into smaller sub-operational periods based on the same ratio to reach timeframes of months, weeks, days, hours, minutes, and even seconds as per the level of accuracy required. Alternatively, prediction for the day can be given based on the planetary position for that day compared with the position in the birth chart.

The Indian (Vedic/Hindu) and Western astrologies differ in various aspects of interpretation. Indian astrology uses zodiac signs based on the position of the birth star and the moon's transit, as compared with the sun's transit which is the concept followed in Western astrology. Each house in the birth chart has its role in interpretation, irrespective of whether it houses a planet. Movement of the planets across the Zodiac as well as planetary and sub-planetary periods are used in interpretation rather than just using transit, in Indian astrology. And the interpretation involves a complex interplay between planets within the house, opposite houses, the ruling planet of the house, as well as the birth star, the current position of the planets, their strengths and weaknesses, and some of the supplementary birth charts (which assess the influence of familial factors, effects of residence, education, spirituality, wealth, etc.) are looked into for prediction. Thereby, this system is comprehensive and sophisticated (Sethi 2007).

There are many other systems of astrology such as Chinese, Mayan, Celtic, Babylonian, etc., which have their own interpretative methods. Debate exists about which of these entities originated first and which later and which was the parent, but one of the critical reviewers of Vedic astrology opines that some elements of Mesopotamian, Greco-Babylonian, and Greco-Egyptian systems were incorporated into the Indian system around the 5th century when Vārāhamihira composed the treatise (Koch 2012/2013). Mayan astrology has a system base of 20, with 20 zodiac signs, one for each day with the cycle repeating every 20 days, 4 determinants based on direction, as well as a second personality zodiac (20 in number) which changes every 13 days in a 260-day calendar. A ruler god / lord of the night for every 9 days and a 52-year cycle indicate a complex method of interpretation, like Vedic astrology (Astrology of the Ancients 2012–2017). The Chinese system on the other hand follows a 60-year cycle with 12 zodiacs represented by animals and 5 elements (air, water, fire, wood, and metal). The day is divided into 12 time periods representing each zodiac animal's prominence. The time and events / omens around the time of birth are used for predictions (Astrosage 2013). Key characteristics of different astrological systems are mentioned in a brief blog article by Segurelha (2008).

### Prediction of Mental Illness Using Indian Astrology

Prediction in Indian astrology involves interplay between various components of the birth chart as well as the planetary movements across the Zodiac. To understand the concept in a simplified way, the birth chart indicates some degree of susceptibility for the illness, while the planetary movements and periods indicate the likelihood of its expression during the time period. With regard to prediction of mental illnesses, the positions of moon, Saturn, and the shadow planets *Rahu–Ketu* play a major role. The ruling planets of the houses, the supporting and opposing nature of planets, their original position at birth, all are taken into consideration while making a prediction. The system also involves a minimum knowledge of Ayurveda for use of terminologies in predicting health and illness. A fair outline about principles used in predicting mental illness can be read in the articles by Tantri (2007) and Chauhan (2014).

### Variants within Indian Astrology

Astrology is not practiced in the same way by all followers of the Indian system. While basic steps in interpretation remain the same, there are two other versions of plotting birth charts that are in vogue. The *Lal Kitaab* uses a fixed-house system no matter where the ascendant is. The interpretation guidelines are modified accordingly to suit simpler analysis and to recommend simple remedies for problems. This came from a mixture of the Vedic and Persian elements of astrology in the 19th Century by Shri Giridhari Lal Sharma in Urdu (Shrimali 2013). The other birth chart plotting method is the *Krishnamurthi Paddathi* (KP system) which appeared in the 1950s and was founded by K. S. Krishnamurthi, a master astrologer from southern India. The KP system incorporates finer aspects of Western astrology and astronomical findings. It takes into account the elliptical pathway of the Earth's revolution which places the cusps at slightly different angles, rendering the zodiac sign's extent of variation from 27° to 33°. In certain scenarios, this can lead to two cusps falling in the same house where the planets in such a house fall into two zodiac signs at the same time, explaining the dual and contradictory effects seen in some horoscopes. This has helped to solve some cases where predictions with traditional Vedic astrology were flawed by simplifying the analysis, thus bypassing the complex and often confusing considerations required when using Vedic astrology (Krishnamurthi 1996, AstroSage no date).

In addition to these birth-chart plotting systems, there are also many variants in analyzing and predicting events based on the already indexed transit charts. These are given by various *Rishis* (sages / saints) and are called *Nādi*. Examples include *Brighu Nandi Nādi, Kashyap Nādi, Hora Nādi, Chandrakala Nādi*, etc. (Pandey 2013). To understand the difference, one can compare *Nādi* to different checklists or reference criteria to make a diagnosis, rate the severity, or predict an outcome. In *Nādi*, a collection of reference birth chart or planetary positions with proven predictions is used for interpretation. However, this study did not involve going beyond the basic principles, and the participating astrologers used only the information provided in the birth chart.

### Courses in Indian Astrology

While there are innumerable courses available (both professional and correspondence) varying from certificate, diploma, degree, to even doctorate, a reliable level of qualification can be the degree course as it enables one to be a teacher in astrology. The bachelor's degree is known as *Vidwan* in *Jyothish* (Scholar / Expert in astrology) and is a three-year course involving theoretical as well as practical training in astrological methods. In addition to the curriculum, students are expected to be knowledgeable in certain Indian scriptures including the Nithi Shastras (law and society matters) and Ayurveda (health and treatment of illnesses), and are trained in spiritual practices / meditation. A degree is awarded once interpretations are consistently accurate, which often involves working as an apprentice under the instructor (personal communication by astrologer Vidwan Shivaprasad Tantri, May 2009).

## Objective

The objective of our study was to determine the predictive ability of Indian astrology (also known as Vedic astrology) in identifying the presence of mental illness based on birth chart and planetary position. We hypothesized that Indian astrology will not be able to predict, that is a Null Hypothesis was used. We aimed at making an objective assessment of the prediction and designed questions for astrologers appropriately. The four predictions relevant to this article include: presence or absence of mental illness, currently ill or healthy, month and year of onset of illness (3-month margin), and core symptoms of illness if any (arbitrary 3-symptom match was taken as agreement).

## Methodology

### *Participants*

This was a prospective cohort study that followed up subjects for 6 months, and was carried out at one of the medical colleges along the west coast of India, a tertiary referral center in the region, between February 2009 and October 2009. The study was approved by the Institutional Ethics Committee of the author's university. Samples consisted of 75 persons with mental illness and 75 age- and sex-matched normal persons (without mental illness). Written informed consent was obtained from all the study participants. The mental illness group included subjects attending inpatient and outpatient care at the Department of Psychiatry, and the normal subjects included relatives of patients as well as volunteers from the community. The astrologers were blind to the study groupings. The inclusion criteria for the subjects included knowledge of their accurate time, date, and place of birth and willingness for regular followup for the following six months. Those unsure about their birth timing, not willing to be followed up, having an organic mental disorder, mental retardation, or in whom diagnosis was unclear were excluded from the study.

## Tools

Socio-demographic and clinical data were collected using a semi-structured pro forma, designed for the study. To make psychiatric diagnoses, the Mini International Neuropsychiatric Interview (MINI) Plus 5.0 was used. The MINI Plus 5.0 is a relatively brief and widely used structured interview for major Axis 1 psychiatric disorders according to DSM IV and ICD 10. It has an acceptably high validity and reliability. It has the facility for investigating substance use disorders and for symptoms due to organic causes (Sheehan, Lecrubier, & Sheehan 1998). The author was trained to use this instrument. the Self-Reporting Questionnaire (SRQ)-24 was used for normal subjects to rule out psychiatric conditions. The SRQ is a self-reported questionnaire developed for the World Health Organization (WHO) to screen mental illness in primary caregivers in developing countries and it also screens for psychotic symptoms (Beusenburg & Orley 1994). The scale has a fairly good sensitivity and specificity at a cutoff of $<7$ for non-psychotic symptoms while the presence of even a single affirmation of psychotic symptoms would mean a need for further evaluation (Chuncholikar 2004). This scale has been validated and is found to be reliable for screening the Indian population (Patel et al. 2008).

## Procedure

Socio-demographic and clinical data were obtained from the study subjects using the semi-structured pro forma. All those with mental illness were evaluated on the MINI Plus 5.0, and the normal subjects on the SRQ. After enrollment, the subjects were assigned a unique number from 1 to 150. To maintain homogeneity in birth-chart plotting, computer-generated birth-chart, operational, and inter-operational periods were plotted using Horoscope Explorer Version 4.0 software (Horoscope Explorer 4.0 2007). The birth charts in this study were plotted on the basic *Vārāhamihira* system as this is the most commonly used plotting system. A sample birth chart with planetary periods used in the study is given in Appendix 1. Each person's birth data, identified now by the allotted random number, was the only information given to the astrologers. All 4 astrologers selected for interpretations in the study had a Qualified Bachelor's Degree in Indian Astrology, following the same *Panchaanga* (Indian calendar) for interpretations, with a minimum of 5 years of experience in interpreting birth charts and planetary position effects, and had publications in the form of astrological books, journal articles, or magazine articles. The astrologers in this study used the principles of Vedic astrology as compiled by Vārāhamihira for interpretation, as it is the most sophisticated and widely used compared with other Indian systems.

**TABLE 1**
**Distribution of Birth Charts to the Astrologers**

| Astrologer | Distribution of birth charts numbered from 1–100 | | Distribution of birth charts numbered from 101–150 | Total birth charts interpreted |
|---|---|---|---|---|
| Astrologer A | Numbers ending with 1, 4, or 7 | + Numbers ending with 0 | Random 8 numbers | 38 + 10 common |
| Astrologer B | Numbers ending with 2, 5, or 8 | + Numbers ending with 0 | Random 7 numbers | 37 + 10 common |
| Astrologer C | Numbers ending with 3, 6, or 9 | + Numbers ending with 0 | Random 7 numbers | 37 + 10 common |
| Astrologer D | | + Numbers ending with 0 | Remaining 28 numbers | 28 + 10 common |

Distribution of birth charts was done as shown in Table 1. Ten birth charts were given to all astrologers to check for inter-astrologer agreement.

Interpretations of the astrologers were matched with actual clinical findings. The terminologies used to describe symptoms and illnesses by the astrologers were from Ayurveda (the ancient Indian medical system using plants and naturally available products for treatment) as well as common words to describe the same in Kannada (a language written and spoken in Southern India). Hence a textbook on Ayurveda written in English and authored by Udupa (2004) and a Kannada-to-English dictionary (Bharadvaj 2005) were used, which gave parallel meanings in allopathic medical practice and English for these terminologies.

### *Statistical Analysis*

Data analysis was done using SPSS for Windows Version 13. Statistical testing was done with the hypothesis being taken as an agreement problem, and a Kappa coefficient was used to test significance. Inter-astrologer agreement was tested using weighted Kappa (Viera & Garette 2005). In addition, interpretation was done viewing astrology as a diagnostic test and analyzed in terms of sensitivity and specificity.

### Results

272 subjects with mental illness and 120 without mental illness were screened; of them 197 persons with mental illness and 45 normal subjects

**TABLE 2**
**Distribution of Socio Demographic and Clinical Variables**

| Variables | Mental illness group $N = 75$ (%) | Normal group $N = 75$ (%) | Statistics |
|---|---|---|---|
| **Mean age** | 32.37 (±12.9) | 32.52 (±12.6) | $t = .070, df = 148$ $p = .944$ |
| **Gender-** | | | |
| Male | 52 (69.3%) | 48 (64%) | $X^2 = .480, df = 1$ |
| Female | 23 (30.7%) | 27 (36%) | $p = .488$ |
| **Clinical Diagnosis (MINI Plus)** | | | |
| Substance use | 2 (2.7%) | | |
| Psychosis | 15 (20%) | | |
| Depression/adjustment | 24 (32%) | — | |
| Mania | 17 (22.7%) | | |
| Anxiety disorders | 16 (21.3%) | | |
| Others | 1 (1.3%) | | |
| **Mean GAF score** | | | |
| At intake | 42.88 (±16.3) | — | |
| **Mean SRQ score** | | | |
| At intake | — | 1.05 (±1.4) | |
| **Strength of belief in astrology** | | | |
| 0–25% | 25 (33.3%) | 12 (16%) | $\chi^2 = 10.487, df = 3$ |
| 26–50% | 13 (17.3%) | 14 (18.7%) | $p = .015*$ |
| 51–75% | 9 (12%) | 4 (5.3%) | |
| 76–100% | 28 (37.3%) | 45 (60%) | |

$*p < 0.05$

declined participation, due to lack of interest, fears regarding confidentiality, magico–religious concerns, or because they could not provide reliable details regarding their date, place, and time of birth. A total sample of 150 individuals was recruited for the study, with 75 members having some mental illness and 75 with no mental illness.

Table 2 gives the characteristics of the two groups at intake. The groups were comparable with respect to gender and to age (mean age 32.37 (±12.9) in the mental illness group and 32.52 (±12.6) in the group without mental illness). Two-thirds of the sample were males; 52 (69.3%) of the

**TABLE 3**
**Agreement Analysis between Clinical and Astrological Findings Regarding Any Mental Illness and Current Mental Illness ($N$ = 150)**

| Variable | Astrology yes | Astrology no | Kappa value | Significance |
|---|---|---|---|---|
| **Life time mental illness** | | | | |
| Clinical yes | 58 | 17 | 0.560 | $p = 0.001$* |
| Clinical no | 16 | 59 | | |
| **Current mental illness** | | | | |
| Clinical yes | 59 | 14 | 0.626 | $p = 0.001$* |
| Clinical no | 14 | 63 | | |

*$p < 0.05$

75 in the mental illness group and 48 (64%) in the without mental illness group. Distribution of diagnoses in the mental illness group ($N$ = 75) were as follows: Commonest diagnosis was of depression/adjustment disorder 24 (32%), followed by mania 17 (22.7%), anxiety disorders 16 (21.3%), psychosis 15 (20%), substance use 2 (2.7%), and others 1 (1.3%). Mean SRQ score in the normal subjects group was 1.05 (SD 1.4). A significant difference was observed between the two groups in terms of their belief regarding astrology ($\chi^2$ = 10.487, $p$ = 0.015).

The agreement rates using the Kappa coefficient were calculated for the clinical impression and the astrological prediction about lifetime occurrence of mental illness and present existence of mental illness. A moderate agreement in the prediction of presence of mental illness ($k$ = 0.560, $p$ = .001) and a substantial agreement in predicting the current state of illness ($k$ = 0.626, $p$ = .001) was found (Table 3).

The inter-astrologer agreement done using 10 birth charts (clinically 4 had mental illness and 6 were normal subjects), found a fair agreement between the astrologers in predicting the presence of mental illness (weighted Kappa 0.2667, significance 0.019*). A non-significant negative agreement (weighted Kappa –0.111, significance 0.805) was found among them for predicting current illness (Table 4).

The comparison of symptom clusters generated in the clinical interview with the ones predicted by the astrologers showed a match of only 26 out of 75 (34.6%) in the mental illness group and 32 out of 75 (42.6%) in the no mental illness group. In the no mental illness group, the symptoms

**TABLE 4**
**Inter-Astrologer Agreement ($N = 10$)**

| Variable | Weighted Kappa value | Significance |
| --- | --- | --- |
| **Presence of mental illness** | 0.2667 | 0.019* |
| **Currently ill** | −0.111 | 0.805 |

*$p < 0.05$

mentioned in the SRQ were matched with ones mentioned by astrologers. The predictions for the time of onset of symptoms (even with a 3-month margin) were also a poor match with 21 out of 75 (28%). Inter-astrologer agreement for onset of illness and symptom cluster could not be measured due to inadequate data.

While viewing astrology as a diagnostic test, it shows a fairly good sensitivity as well as specificity in identifying lifetime mental illness (>75%) as well as presence or absence of mental illness at the current time (>80%). The ability of astrologers to correctly predict that a person is suffering from mental illness (positive predictive value) as well as the ability to predict that a person is not affected with a mental illness (negative predictive value) is also promising in both the contexts (Table 5).

## Discussion

The primary findings in the study suggest that astrology as a tool is a fairly good predictor of mental illness. This study found a moderate agreement in the prediction of the presence of mental illness and substantial agreement with regard to prediction of the current state of illness. The fairly high degree of sensitivity as well as specificity when astrology is viewed as a diagnostic test indicates that the birth chart could possibly be looked at for these predictions. The lifetime risk is basically derived by the astrologers based on the planetary positions in the birth chart. The current status of illness is usually interpreted based on the current planetary position and planetary sub-periods (called *dashas* and *antardashas* in Indian astrology) where unfavorable planets would result in symptoms (Tantri 2007, Chauhan 2014). Studies involving the birth chart using Western astrology that looked at personality profiles found no more than chance association in the predictive ability of the astrologers (Carlson 1985, Wyman & Vyse

**TABLE 5**
**Predictive Validity of Astrology as a Diagnostic Test**

| Variable | Sensitivity | Specificity | Positive predictive value | Negative predictive value | Area under ROC curve* |
|---|---|---|---|---|---|
| **Life-time mental illness** | 77.3% | 78.7% | 78.4% | 77.6% | 0.781 |
| **Current illness** | 80.8% | 81.8% | 80.8% | 81.8% | 0.813 |

* Area = 1 would indicate ideal test with 100% sensitivity and 100% specificity.

2008). One study that looked at the association of certain planets and sun signs with schizophrenia gave only partial support to astrological principles (Ohaeri 1997). Among the studies done so far, only Carlson's experiment (1985) had involved astrologers and had found that the astrologer's claim of even 50% accuracy was an exaggeration. The results had actually shown one-third accuracy. A reanalysis of the same data in the appraisal given by Ertel (2009) had actually shown a small significance and opined the study analysis to be biased and inclined toward proving the preset agenda of astrology as bogus. In contrast, the current study has found a moderate agreement which is possibly the best obtained through research so far. This study also addressed Ertel's (2009) criticism of Carlson's unfair study design by having a two-choice discrimination (yes/no) for the questions on lifetime and current mental illness. However, this finding needs to be interpreted with caution because a subsequent analysis showed that inter-astrologer agreement (among the participating astrologers) was not perfect. Hence there is still a possibility of this being a chance association.

On the other hand, astrology clearly failed to meet agreement criteria in the onset of illness and the symptom cluster match. The failure of indicating the correct time of onset, but with matching of the current mental status, raises the doubt that it could be just a chance association. If one could tell the current status based on the planetary periods, then one should ideally be able to correctly predict the onset. However, from the astrologer's viewpoint, the accuracy of a prediction depends on the depth of analysis made. Planetary periods vary from 6 years to 20 years, and the 1st level of planetary subperiods vary from a few months to a few years. Each of these subperiods can be analyzed with further subperiods until it can come down to minutes and seconds. Each subperiod signifies fluctuations within an illness period based on whether the subperiod for that individual is favorable or adverse. Such an extent of analysis would invariably require a lot of time and is practically not feasible (personal communication Shivaprasad Tantri, 2009; Sethi 2007). This could explain the inter-astrologer difference when two of the astrologers went further in their analysis for predictions, while the other two stuck to the information on the chart provided. The same argument could also explain why the astrologers failed to pinpoint the onset of illness while still being able to predict more accurately whether the individual was currently ill. While this argument may still be valid, with the current data it can only be stated that onset could not be predicted accurately by the system used by the participating astrologers.

The analysis of predictions like the symptom cluster often puts across possibilities of bias due to the conceptual difference in the classification of illness in the two systems. A person with an inclination / belief in astrology is likely to interpret it as an identifying feature, whereas others may disagree. A similar criticism was made by Crowe (1990), who said that people tend to internalize astrological reports when given in vague terms and then accept them as real. The author attempted to minimize that impact by keeping the astrologer and the subjects blind to each other, and also referred to an Ayurveda textbook (as terminologies used in astrology to describe illnesses are the same) that links illnesses with allopathic classifications. However, the open-ended nature of the question, the non-specific nature of some of the psychiatric symptoms, the vagueness with which some of the terminologies are used by the astrologers, and also the arbitrary criteria of the three-symptom match itself, questions the method adopted to test it. Getting an exact fit for diagnostic categories between allopathic classification and astrological classification is a challenge that needs to be dealt with to draw reliable conclusions. Alternatively, a diagnosis made as per Ayurveda might be a better choice to objectively test.

The present study had 4 astrologers with similar theoretical backgrounds predicting, hence it may be expected that their agreement would be good. From the results, the inter-astrologer agreement was not good except for prediction of lifetime occurrence of mental illness. This type of inconsistency brings out the vagueness involved in the interpretation and is highlighted in the Carlson study (1985). However, the sample size used for inter-astrologer agreement was inadequate (10 (7%), $N = 150$). Ideally, to derive a firm conclusion about inter-astrologer agreement, the sample size required would be the same as the study when analyzing it as an agreement problem. This limitation allays the decision to accept or reject the original study hypothesis.

The sample size in our study was comparable with previous studies mentioned in the Introduction. The subjects in the mental illness group were comparable with the subjects in the no mental illness group in terms of age and gender. The distribution of diagnoses was representative of the usual pattern seen in the hospital where the study was conducted. However, there was a relative paucity of substance-use disorders in the sample. This might not be relevant, as the present study, by design, addressed the larger issue of mental illness versus normalcy more specifically than individual diagnoses. The belief in astrology was higher in the normal group as compared with the mental illness group. This could be because of the hospital catchment area being larger where the belief might not be strong. The higher belief in the local population is reflective of the persisting trend seen from the study by Kapur (1975).

The blinding of astrologers as well as of the subjects makes the study unique. In the previous studies using birth charts (Western astrology), there had always been contact between subjects and astrologers or their interpretations for checking accuracy. Both would be adding to the bias (Carlson 1985, Wyman & Vyse 2008). Furthermore, it needs to be noted that none of the astrologers took any remuneration for their work and being academically oriented believed in scientific testing of their predictions. Their participation was of their own free will and out of curiosity to know the results of this investigation untainted by expectation of pecuniary benefits.

The study finds strength in its adequate sample size, testing using a comprehensive system of astrology rather than just components, blinding of the subjects and astrologers, and designing it to assess objectively at least some of the predictions. Astrological analysis can be conceptualized under prediction of the state / event and the precision of its occurrence. This study objectively tested the former but could not control the latter. We recommend the use of the Indian astrology system based on the principles of Vārāhamihira (as a comprehensive system) for future studies rather than limited and piecemeal interpreting systems. The limitations that need to be worked on include: having a single astrologer or ones with good inter-astrologer agreement, a common disease concept or a study design for objectively assessing symptoms and illness, a defined depth of astrological analysis, and a better tool for capturing lifetime mental illness in the 'normals' group.

### Conclusion

A sizable number of people believe in astrology despite controversies regarding its veracity, and these approaches tend to be sought and practiced in *pari passu* with the use of allopathic psychiatric treatments in the community. As a parallel system of thought, it continues to interest researchers applying modern scientific methods to evaluate it. Theoretical differences in conceptualizing mental illness continue to produce difficulties in scientifically testing all aspects of prediction by astrological means. Overall, on completion of the present exercise, the evidence seems to point toward astrology as practiced in the study modestly predicting the presence of mental illness. Future studies can standardize the analysis to test the precision of occurrences.

## Acknowledgments

We thank the 4 astrologers for participating in the study and doing the interpretations out of academic interest without seeking any remuneration.

**Funding:** The authors received no financial support for the research, authorship, and/or publication of this article.

## References

Abdel-Khalek A., & Lester D. (2006). Astrological signs and personality in Kuwaitis and Americans. *Psychological Reports, 98*:602–607.

Astrology of the Ancients (2012–2017). Mayan Astrology. Well Read Gnome [Blog Post]. http://www.astrologyoftheancients.com/mayan-astrology/

Astrosage (2013). Difference Between Vedic Astrology & Chinese Astrology. November 26. http://astrology.astrosage.com/2013/11/difference-between-vedic-astrology.html

Astrosage (no date). K. P. System: An Introduction. http://www.astrosage.com/kpSystemIntro.asp

Barr, W. (2000). Lunacy revisited. The influence of the moon on mental health and quality of life. *Journal of Psychosocial Nursing Mental Health Services, 38*:28–35.

Baxendale, S., & Fisher J. (2008). Moonstruck? The effect of the lunar cycle on seizures. *Epilepsy Behavior, 13*:549–550.

Benbadis, S. R., Chang, S., Hunter J., & Wang, W. (2004). The influence of the full moon on seizure frequency: Myth or reality? *Epilepsy Behavior, 5*:596–597.

Beusenburg, M., & Orley, J. (1994). *A User's Guide to Self-Reporting Questionnaire (SRQ).* Geneva: World Health Organization.

Bharadvaj, D. K. (2005). *Standard Kannada English Dictionary.* Gadag, India: BGS Publications.

Carlson, S. (1985). A double-blind test of astrology. *Nature, 318*:419–425.

Chauhan, A. (2014). Predicting psychological disorders by astrology. *Journal of Psychotherapy and Psychological Disorders, 2*:1. DOI:10.4172/2327-4654.1000103

Chuncholikar, S. V. (2004). Use of SRQ in psychiatric epidemiology. *Indian Journal Community Medicine, 29*:190–191.

Crowe, R. A. (1990). Astrology and the scientific method. *Psychology Reports, 67*:163–191.

Ertel, S. (2009). Appraisal of Shawn Carlson's renowned astrology tests. *Journal of Scientific Exploration, 23*(2):125–137.

Hare, E. H. (1977). Medical astrology and its relation to modern psychiatry. *Proceedings of the Royal Society of Medicine, 70*:105–110.

Hicks-Caskey, W. E., & Potter, D. R. (1991). Effect of the full moon on a sample of developmentally delayed, institutionalized women. *Perceptual and Motor Skills, 72*:1375–1380.

Horoscope Explorer 4.0. (2007). CD from Public Soft. http://www.itbx.com

Kapur, R. L. (1975). Mental health care in rural India: A study of existing patterns and their implications for future policy. *British Journal of Psychiatry, 127*:286–293.

Koch, D. (2012/2013). Vedic Astrology—Critically examined.
http://www.astro.com/astrology/in_vedic2_e.htm

Krishnamuthi, K. S. (1996). Stellar Astrological—Reader no. V (Transit—*Gocharapala Nirnayam*). Seventh Edition. Madras/Chennai, India: Krishman & Co.

Ohaeri, J. U. (1997). The planetary positions and relationships at the dates of birth of a cohort of Nigerian schizophrenics. *African Journal of Medicine and Medical Sciences, 26*:127–133.

Owen, C., Tarantello, C., Jones, M., & Tennant, C. (1998). Lunar cycles and violent behaviour. *Australian and New Zealand Journal of Psychiatry, 32*:496–499.

Pandey, P. (2013, February 11). What is Nadi Astrology?
http://www.astrosage.com/nadiastrology/

Patel, V., Araya, R., Chowdhary, N., et al. (2008). Detecting common mental disorders in primary care in India: A comparison of five screening questionnaires. *Psychological Medicine, 38*:221–228.

Polychronopoulos, P., Argyriou, A. A., Sirrou, V., Huliara, V., Aplada, M., & Gourzis P. (2006). Lunar phases and seizure occurrence: Just an ancient legend? *Neurology, 66*:1442–1443.

Segurelha (2008, July 2). Mayan Astrology, Celtic Astrology, Chinese Astrology, Vedic Astrology, Sideral Astrology: Which is Correct?
http://astrotransits.blogspot.in/2008/07/mayan-astrology-celtic-astrology.html

Sethi, A. (2007). *Learn Astrology in 30 Days.* New Delhi, India: Computer Zone Publishers.

Sharma, R. N., & Thakur, C. P. (1980). Full moon and poisoning. *British Medical Journal, 281*:1684.

Sharma, B. K., Prasad, P. V., & Narayana A. (2007). Contribution of astrology in medicine—A review. *Bulletin of the Indian Institute of History of Medicine, Hyderabad, 37*:45–62.

Sharma, B. K., Subhakta, P. K., & Narayana A. (2007). Health promotion, preventive and curative aspects of diseases in astrology. *Bulletin of the Indian Institute of History of Medicine, Hyderabad, 37*:135–152.

Sheehan, D., Lecrubier, Y., & Sheehan, K. (1998). The Mini International Neuropsychiatric Interview (MINI): The development and validation of a structured diagnostic psychiatric interview for DSM-IV and ICD-10. *Journal of Clinical Psychiatry, 20*:22–33.

Shrimali, R. (2013). *Lal Kitab (Tested Remedies of the Famous and Unique Urdu Astrological Treatise Lal Kitab).* New Delhi, India: Diamond Pocket Books.

Snelson, A. (2004). Under the Brighton full moon. *Mental Health Practice, 8*:30-34.

Tantri, S. (2007). Predicting illness with Astrology [title in Kannada language "Jyothishyaastradalli Roga Chinthane"]. *Yoga Deepika,* 36–39.

Udupa, M. (2004). *Comprehensive Kayachikitsa and Principles of Ayurveda.* First edition. Bangalore: Laveena Publications.

Viera, A. J., & Garette, J. M. (2005). Understanding inter-observer agreement: The kappa statistic. *Family Medicine, 37*:360–363.

Wyman, A. J., & Vyse, S. (2008). Science versus the stars: A double-blind test of the validity of the NEO Five-Factor Inventory and computer-generated astrological natal charts. *Journal of General Psychology, 135*:287–300.

Zargar, M., Khaji, A., Kaviani, A., Karbakhsh, M., Yunesian, M., & Abdollahi, M. (2004). The full moon and admission to emergency rooms. *Indian Journal of Medical Sciences, 58*:191–195.

Zimecki, M. (2006). The lunar cycle: Effects on human and animal behavior and physiology. *Postepy Higieny i Medycyny Doswiadczalnej* [Advances in Hygiene and Experimental Medicine], *60*:1–7. https://www.ncbi.nlm.nih.gov/pubmed/16407788#

# APPENDIX 1

## Sample Horoscope Data Provided to the Astrologers

See 7 horoscope charts below.

**Appendix 1**                                                                  1

| | |
|---|---|
| Date of Birth: | 12 November 1976 Friday |
| Time of Birth: | 05:43 AM |
| Place of Birth: | Bangalore (Karnataka), India |
| Latitude: | 13.0N   Longitude:  77.35E |
| Ayanmash: | NC Lahiri        23:32:0 |
| Local Mean Time: | 05:23:20 |
| Sidreal Time: | 8:48:20 |
| LT Correction: | -19:40 |
| Obliq: | 23.4 |

### Avkahada Chakra

| | |
|---|---|
| **Lagna** | Libra |
| **Lagna Lord** | Venus |
| **Rashi** | Gemini |
| **Rashi Lord** | Mercury |
| **Nakshatra** | Punarvasu |
| **Nakshatra Lord** | Jupiter |
| **Charan** | 1 |
| **Tithi** | Panchami Krishna |
| **Paya** | Silver |
| **S.S. Yoga** | Sadhya |
| **Karan** | Taitika |
| **Varna** | Vaishya |
| **Tatwa** | Jala |
| **Vashya** | Manav |
| **Yoni** | Cat(F) |
| **Gana** | Deva |
| **Nadi** | Aadi |
| **Nadi Pada** | Aadi |
| **Vihaga** | Pingala |
| **First Letters** | Kay, Ko, Haa, Hee |
| **Sun Sign** | Scorpio |
| **Decanate** | 2 |

### Ghatak(Malefics)

| | |
|---|---|
| **Rashi** | Kumbh |
| **Months** | Ashad |
| **Tithi** | 2, 7, 12 |
| **Day** | Monday |
| **Nakshatra** | Swati |
| **Prahar** | 3 |
| **Lagna** | Kark |
| **Yoga** | Parigha |
| **Karan** | Kaulava |

### Favourable Points

| | |
|---|---|
| **Lucky Numbers** | 3 |
| **Good Numbers** | 1, 2, 3, 9 |
| **Evil Numbers** | 4, 5, 8 |
| **Good Years** | 12,21,30,39,48 |
| **Lucky Days** | Sat, Wed, Fri |
| **Good Planets** | Sat, Merc, Ven |
| **Evil Planets** | Mars, Jupiter |
| **Friendly Signs** | Gem Cap Aqua |
| **Good Lagna** | Cap, Ari, Gem, Leo |
| **Lucky Metal** | Silver |
| **Lucky Stone** | Diamond |
| **Lucky Time** | Sunrise |
| **Lucky Direction** | South-East |

**Public Software Library India Pvt Ltd - 263, Sant Nagar, New Delhi-110065, Email: psl@nde.vsnl.net.in**

Generated by Horoscope Explorer (c) Public Software Library India Pvt Ltd - www.itbix.com

Appendix **1**  **2**

## Planetary Positions at Birth Time

| Planets | Rashi | Degrees | Nakshatra | Nak Lord | Pad | Dir | Dignity |
|---------|-------|---------|-----------|----------|-----|-----|---------|
| Ascendant | Libra | 17:24:06 | Swati | Rahu | 4 | | - |
| Sun | Libra | 26:12:24 | Vishakha | Jupiter | 2 | Direct | - |
| Mercury | Libra | 28:56:15 | Vishakha | Jupiter | 3 | Direct | - |
| Venus | Sagittarius | 03:31:25 | Moola | Ketu | 2 | Direct | - |
| Mars | Scorpio | 00:05:03 | Vishakha | Jupiter | 4 | Direct | Own House |
| Jupiter | Taurus | 03:27:04 | Krittika | Sun | 3 | Retro | - |
| Saturn | Cancer | 23:06:40 | Ashlesha | Mercury | 2 | Direct | - |
| Moon | Gemini | 21:39:00 | Punarvasu | Jupiter | 1 | Direct | - |
| Rahu | Libra | 08:59:56 | Swati | Rahu | 1 | Retro | - |
| Ketu | Aries | 08:59:56 | Ashwini | Ketu | 3 | Retro | - |
| Uranus | Libra | 14:42:35 | Swati | Rahu | 3 | Direct | - |
| Neptune | Scorpio | 19:16:44 | Jyestha | Mercury | 1 | Direct | - |
| Pluto | Virgo | 19:28:52 | Hasta | Moon | 3 | Direct | - |

## Lagna Kundali

| 12 | 1 Ket | 2 Jup | 3 Moon |
|----|-------|-------|--------|
| 11 | | | 4 Sat |
| 10 | | | 5 |
| 9 Ven | 8 Mar Nep | 7 Asc Sun Mer Rah Ura | 6 Plu |

## Saptamamsha — Children

| 12 | 1 Sun Mer | 2 Mar | 3 Sat Ket |
|---|---|---|---|
| 11 Asc | | | 4 Plu |
| 10 Ura | | | 5 |
| 9 Ven Rah | 8 Jup Moon | 7 | 6 Nep |

## Navamsha — Spouse

| 12 Asc | 1 Moon | 2 Sun Ven | 3 Mer Ket Plu |
|---|---|---|---|
| 11 Jup Ura | | | 4 Mar |
| 10 Sat | | | 5 |
| 9 Rah Nep | 8 | 7 | 6 |

## Dashamamsha — Profession

| 12 Asc | 1 | 2 | 3 Sun Ket |
|---|---|---|---|
| 11 Jup Ura | | | 4 Mer Mar |
| 10 Ven Moon Nep | | | 5 |
| 9 Rah | 8 Plu | 7 Sat | 6 |

## Dwadashamamsha — Parents

| 12 Ura | 1 Asc Sat Plu | 2 | 3 Jup Nep |
|---|---|---|---|
| 11 Moon | | | 4 Ket |
| 10 Ven Rah | | | 5 Sun |
| 9 | 8 Mar | 7 | 6 Mer |

Appendix 1  4

## Bhav Table

| Bhav | Bhav Begins | | Mid Bhav | |
|------|-------------|---------|-------------|---------|
| 1 | Libra | 02:11:10 | Libra | 17:24:06 |
| 2 | Scorpio | 02:11:10 | Scorpio | 16:58:15 |
| 3 | Sagittarius | 01:45:19 | Sagittarius | 16:32:23 |
| 4 | Capricorn | 01:19:27 | Capricorn | 16:06:31 |
| 5 | Aquarius | 01:19:27 | Aquarius | 16:32:23 |
| 6 | Pisces | 01:45:19 | Pisces | 16:58:15 |
| 7 | Aries | 02:11:10 | Aries | 17:24:06 |
| 8 | Taurus | 02:11:10 | Taurus | 16:58:15 |
| 9 | Gemini | 01:45:19 | Gemini | 16:32:23 |
| 10 | Cancer | 01:19:27 | Cancer | 16:06:31 |
| 11 | Leo | 01:19:27 | Leo | 16:32:23 |
| 12 | Virgo | 01:45:19 | Virgo | 16:58:15 |

### Bhav Chalit Chakra

| 12 | 1 Ket | 2 Jup | 3 Moon |
|----|-------|-------|--------|
| 11 | | | 4 Sat |
| 10 | | | 5 |
| 9 Ven | 8 Nep | 7 Asc Sun Mer Mar Rah Ura | 6 Plu |

### Chandra Kundli

| 12 | 1 Ket | 2 Jup | 3 Moon |
|----|-------|-------|--------|
| 11 | | | 4 Sat |
| 10 | | | 5 |
| 9 Ven | 8 Mar Nep | 7 Asc Sun Mer Rah Ura | 6 Plu |

## Sun Based Upagrahas

| Upagraha | Lord | Rashi | Degrees | Nakshatra | Charan |
|---|---|---|---|---|---|
| Dhuma | Mars | Pisces | 09:32:24 | U.Bhadra | 2 |
| Vyatipata | Rahu | Aries | 20:27:36 | Bharani | 3 |
| Parivesha | Moon | Libra | 20:27:36 | Vishakha | 1 |
| Indrachapa | Venus | Virgo | 09:32:24 | Uttara | 4 |
| Upaketu | Ketu | Virgo | 26:12:24 | Chitra | 1 |
| Bhukampa | | Capricorn | 16:12:24 | Sravana | 2 |
| Ulka | | Aquarius | 26:12:24 | P.Bhadra | 2 |
| Brahmadanda | | Taurus | 02:52:24 | Krittika | 2 |
| Dhwaja | | Cancer | 22:52:24 | Ashlesha | 2 |

## Weekday Based Upagrahas (Parashara)

| Upagraha | Rashi | Degrees | Lord | Nakshatra | Charan |
|---|---|---|---|---|---|
| Kalabela | Sagittarius | 01:50:23 | Jupiter | Moola | 1 |
| Paridhi | Capricorn | 15:32:19 | Saturn | Sravana | 2 |
| Mrityu | Aries | 27:16:26 | Mars | Krittika | 1 |
| Ardhaprahara | Gemini | 10:59:22 | Mercury | Aridra | 2 |
| Yamakantaka | Cancer | 22:09:40 | Moon | Ashlesha | 2 |
| Kodanda | Virgo | 06:53:54 | Mercury | Uttara | 4 |
| Gulika | Libra | 20:39:54 | Venus | Vishakha | 1 |
| Mandi | Scorpio | 23:39:26 | Mars | Jyestha | 3 |

## Weekday Based Upagrahas (Kalidasa)

| Upagraha | Rashi | Degrees | Lord | Nakshatra | Charan |
|---|---|---|---|---|---|
| Kalabela | Capricorn | 15:32:19 | Saturn | Sravana | 2 |
| Paridhi | Aries | 27:16:26 | Mars | Krittika | 1 |
| Mrityu | Gemini | 10:59:22 | Mercury | Aridra | 2 |
| Ardhaprahara | Cancer | 22:09:40 | Moon | Ashlesha | 2 |
| Yamakantaka | Virgo | 06:53:54 | Mercury | Uttara | 4 |
| Kodanda | Libra | 20:39:54 | Venus | Vishakha | 1 |
| Gulika | Sagittarius | 01:50:23 | Jupiter | Moola | 1 |
| Mandi | Scorpio | 23:39:26 | Mars | Jyestha | 3 |

Appendix 1  6

## Vimshottari Dasha (Mahadasha)

| Jupiter | 16 Years | | Saturn | 19 Years | | Mercury | 17 Years |
|---|---|---|---|---|---|---|---|
| From | 19-11-1974 | | From | 19-11-1990 | | From | 19-11-2009 |
| To | 19-11-1990 | | To | 19-11-2009 | | To | 19-11-2026 |
| Jupiter | 19-11-1974 | | Saturn | 19-11-1990 | | Mercury | 19-11-2009 |
| Saturn | 07-01-1977 | | Mercury | 22-11-1993 | | Ketu | 16-04-2012 |
| Mercury | 21-07-1979 | | Ketu | 01-08-1996 | | Venus | 13-04-2013 |
| Ketu | 26-10-1981 | | Venus | 09-09-1997 | | Sun | 12-02-2016 |
| Venus | 02-10-1982 | | Sun | 09-11-2000 | | Moon | 18-12-2016 |
| Sun | 02-06-1985 | | Moon | 22-10-2001 | | Mars | 19-05-2018 |
| Moon | 22-03-1986 | | Mars | 23-05-2003 | | Rahu | 17-05-2019 |
| Mars | 21-07-1987 | | Rahu | 02-07-2004 | | Jupiter | 04-12-2021 |
| Rahu | 26-06-1988 | | Jupiter | 08-05-2007 | | Saturn | 10-03-2024 |

| Ketu | 7 Years | | Venus | 20 Years | | Sun | 6 Years |
|---|---|---|---|---|---|---|---|
| From | 19-11-2026 | | From | 19-11-2033 | | From | 19-11-2053 |
| To | 19-11-2033 | | To | 19-11-2053 | | To | 19-11-2059 |
| Ketu | 19-11-2026 | | Venus | 19-11-2033 | | Sun | 19-11-2053 |
| Venus | 17-04-2027 | | Sun | 21-03-2037 | | Moon | 09-03-2054 |
| Sun | 17-06-2028 | | Moon | 21-03-2038 | | Mars | 07-09-2054 |
| Moon | 23-10-2028 | | Mars | 19-11-2039 | | Rahu | 13-01-2055 |
| Mars | 24-05-2029 | | Rahu | 19-01-2041 | | Jupiter | 08-12-2055 |
| Rahu | 20-10-2029 | | Jupiter | 19-01-2044 | | Saturn | 25-09-2056 |
| Jupiter | 07-11-2030 | | Saturn | 20-09-2046 | | Mercury | 07-09-2057 |
| Saturn | 14-10-2031 | | Mercury | 20-11-2049 | | Ketu | 14-07-2058 |
| Mercury | 23-11-2032 | | Ketu | 19-09-2052 | | Venus | 19-11-2058 |

| Moon | 10 Years | | Mars | 7 Years | | Rahu | 18 Years |
|---|---|---|---|---|---|---|---|
| From | 19-11-2059 | | From | 19-11-2069 | | From | 19-11-2076 |
| To | 19-11-2069 | | To | 19-11-2076 | | To | 19-11-2094 |
| Moon | 19-11-2059 | | Mars | 19-11-2069 | | Rahu | 19-11-2076 |
| Mars | 18-09-2060 | | Rahu | 17-04-2070 | | Jupiter | 02-08-2079 |
| Rahu | 19-04-2061 | | Jupiter | 05-05-2071 | | Saturn | 26-12-2081 |
| Jupiter | 19-10-2062 | | Saturn | 10-04-2072 | | Mercury | 31-10-2084 |
| Saturn | 18-02-2064 | | Mercury | 20-05-2073 | | Ketu | 20-05-2087 |
| Mercury | 19-09-2065 | | Ketu | 17-05-2074 | | Venus | 07-06-2088 |
| Ketu | 18-02-2067 | | Venus | 13-10-2074 | | Sun | 07-06-2091 |
| Venus | 19-09-2067 | | Sun | 13-12-2075 | | Moon | 01-05-2092 |
| Sun | 21-05-2069 | | Moon | 19-04-2076 | | Mars | 31-10-2093 |

## Vimshottari-Pratyantar

| Saturn | | Mercury | | Ketu | |
|---|---|---|---|---|---|
| From | 19-11-1990 | From | 22-11-1993 | From | 01-08-1996 |
| To | 22-11-1993 | To | 01-08-1996 | To | 09-09-1997 |
| Saturn | 19-11-1990 | Mercury | 22-11-1993 | Ketu | 01-08-1996 |
| Mercury | 12-05-1991 | Ketu | 10-04-1994 | Venus | 24-08-1996 |
| Ketu | 15-10-1991 | Venus | 07-06-1994 | Sun | 31-10-1996 |
| Venus | 18-12-1991 | Sun | 18-11-1994 | Moon | 20-11-1996 |
| Sun | 18-06-1992 | Moon | 06-01-1995 | Mars | 24-12-1996 |
| Moon | 12-08-1992 | Mars | 29-03-1995 | Rahu | 16-01-1997 |
| Mars | 11-11-1992 | Rahu | 25-05-1995 | Jupiter | 18-03-1997 |
| Rahu | 15-01-1993 | Jupiter | 19-10-1995 | Saturn | 11-05-1997 |
| Jupiter | 28-06-1993 | Saturn | 28-02-1996 | Mercury | 14-07-1997 |

| Venus | | Sun | | Moon | |
|---|---|---|---|---|---|
| From | 09-09-1997 | From | 09-11-2000 | From | 22-10-2001 |
| To | 09-11-2000 | To | 22-10-2001 | To | 23-05-2003 |
| Venus | 09-09-1997 | Sun | 09-11-2000 | Moon | 22-10-2001 |
| Sun | 21-03-1998 | Moon | 27-11-2000 | Mars | 09-12-2001 |
| Moon | 18-05-1998 | Mars | 25-12-2000 | Rahu | 12-01-2002 |
| Mars | 22-08-1998 | Rahu | 15-01-2001 | Jupiter | 09-04-2002 |
| Rahu | 29-10-1998 | Jupiter | 08-03-2001 | Saturn | 25-06-2002 |
| Jupiter | 20-04-1999 | Saturn | 23-04-2001 | Mercury | 25-09-2002 |
| Saturn | 21-09-1999 | Mercury | 17-06-2001 | Ketu | 15-12-2002 |
| Mercury | 23-03-2000 | Ketu | 05-08-2001 | Venus | 18-01-2003 |
| Ketu | 02-09-2000 | Venus | 25-08-2001 | Sun | 25-04-2003 |

| Mars | | Rahu | | Jupiter | |
|---|---|---|---|---|---|
| From | 23-05-2003 | From | 02-07-2004 | From | 08-05-2007 |
| To | 02-07-2004 | To | 08-05-2007 | To | 19-11-2009 |
| Mars | 23-05-2003 | Rahu | 02-07-2004 | Jupiter | 08-05-2007 |
| Rahu | 16-06-2003 | Jupiter | 05-12-2004 | Saturn | 09-09-2007 |
| Jupiter | 16-08-2003 | Saturn | 23-04-2005 | Mercury | 02-02-2008 |
| Saturn | 09-10-2003 | Mercury | 05-10-2005 | Ketu | 12-06-2008 |
| Mercury | 12-12-2003 | Ketu | 01-03-2006 | Venus | 05-08-2008 |
| Ketu | 07-02-2004 | Venus | 01-05-2006 | Sun | 06-01-2009 |
| Venus | 02-03-2004 | Sun | 21-10-2006 | Moon | 22-02-2009 |
| Sun | 08-05-2004 | Moon | 12-12-2006 | Mars | 10-05-2009 |
| Moon | 28-05-2004 | Mars | 09-03-2007 | Rahu | 03-07-2009 |

*Journal of Scientific Exploration,* Vol. 32, No. 3, pp. 579–595, 2018      0892-3310/18

*COMMENTARY*

# Investigating "Physical Mediums" via Audio Signal Processing: A Comment on a Recent Approach

## Michael Nahm

Institute for Frontier Areas of Psychology and Mental Health (IGPP), Freiburg, Germany
nahm@igpp.de

Submitted: June 5, 2018; Accepted August 3, 2018; Published September 30, 2018

DOI: https://doi.org/10.31275/2018/1334
Creative Commons License CC-BY-NC-ND

**Abstract**—Technological advances in recent decades offer possibilities to study the phenomena of physical mediumship using new approaches of investigation. One new approach is the analyses of supposed "spirit voices" recorded in dark séance rooms during sittings with the alleged physical medium Warren Caylor. These analyses were performed by a professor of applied informatics, Eckhard Kruse. He concluded that the results of his investigations provide evidence, if not proof, demonstrating that these voices are indeed produced by materialized "spirits," and he publicized this interpretation of his work via various dissemination channels. As I will show in this article, however, Kruse's approach to studying these "spirit voices" is loaded with conceptual and methodological deficiencies that ultimately render his claims untenable.

*Keywords:* Physical mediumship—spirit voices—voice analyses—Eckhard
        Kruse—Warren Caylor

## Introduction

In recent years, a German professor of applied informatics, Eckhard Kruse, broadcast the results of voice analyses he performed on "spirit voices" that manifested during the séances of an alleged medium for physical mediumship, Warren Caylor, in articles (Kruse 2016a, 2016b, 2018a), on his website (Kruse 2017a, 2017b, 2018b, 2018c), and in talks and interviews (e.g., Kruse 2016c, 2017c, Maier 2016). Hardly knowing anything about physical mediumship before, he was first introduced to the strange phenomena that happen in that context in spring 2015 by Lucius Werthmüller, the head of the Basel Psi Association, when he visited a public séance of another ostensible physical medium, Kai Mügge (Kruse 2015, 2016a). Kruse became fascinated by what he experienced, and he tried to advance studies into physical mediumship (Kruse 2015, 2017d). With the permission of Werthmüller and the supposed mediums, he introduced technical equipment into séance rooms to document and analyze some of their phenomena. The voices Kruse analyzed were recorded during the séances of Caylor at the Basel Psi Association. For this purpose, Kruse mounted four microphones to a wall at a distance from each other. The recordings of these four microphones allow for the audio-localization of the source of a "spirit voice," and indeed he was able to trace the movements of voices that are thought to belong to fully materialized "spirits" inside the séance room after analyzing the recordings with special software. Moreover, Kruse analyzed these recorded voices further regarding certain characteristics of human voices and their formants. These formants represent amplitude peaks in the frequency spectrum of voices, and they are determined by the individual anatomical makeup of one's vocal tract. Important formants include formants F1–F4, as they are considerably involved in shaping the individual characteristics of the sound of human voices. Kruse believes that the frequencies of F3 and F4 can hardly be altered when people disguise their voice even dramatically, and he claims that their alleged immutability would represent an important feature of "voice forensics." During the séances, he recorded voices that displayed considerable differences in F3 and F4. Consequently, Kruse considers the phenomena of Caylor genuine (e.g., Kruse 2016a, 2016b, Maier 2016), and stated that his measurements of Caylor's "spirit voices" exclude all possibilities of producing them in a fraudulent manner that lead to hypotheses of deception ad absurdum (Maier 2016:179). In his other contributions on this subject, Kruse advanced basically the same opinion, albeit in more or less attenuated formulations.

Moreover, Kruse considers the binding and gagging methods applied to secure Caylor at the Basel Psi Association safe, and he assumes that Caylor always rests bound and gagged on his chair throughout the séances (Kruse 2016b, 2016c, 2017a, 2017b, 2017c). Apparently, he holds much sympathy for Caylor and other alleged mediums, and he doesn't deem it possible that they might be cheating on him, although the history of physical mediumship brims with examples in which researchers were duped by supposedly honest and trustworthy medium-friends (for numerous examples see Gulat-Wellenburg, Klinckowstroem, & Rosenbusch 1925). A famous example of the past concerns the pseudo-medium Ladislaus Laszlo (Schrenck-Notzing 1924), a recent example concerns Kai Mügge (Nahm 2014, 2016).[1]

Kruse's approach to studying séance phenomena using sophisticated modern technologies is innovative and interesting, and clearly it helps to obtain insights into the phenomena occurring in séances held in complete darkness. Nevertheless, as I will show in the present article, Kruse's claim that his investigation provides evidence (or even proof) that the "spirit voices" manifesting at Caylor's séances are genuine, is premature and untenable. Nevertheless, Werthmüller and other uncritical actors sympathizing with parapsychology, physical mediumship, and esoterics, propagate Kruse's claims further in interviews and talks (e.g., Werthmüller 2017). Hence, to counter the growing stream of misinformation to the public, a few comments on the medium Kruse investigated and on the methodological approach he pursued seem apt. In the following, I will first comment on typical control measures applied at the Basel Psi Association to secure ostensible mediums, then on the supposed medium Warren Caylor, and finally, on the crucial feature of Kruse's voice analyses, the determination of voice formants.

### Comment on the Controls Applied at Commercial Sittings with Alleged Physical Mediums at the Basel Psi Association

To begin with, it should be noted that the methods of binding and controlling "mediums" to prevent them from producing fraudulent phenomena at the Basel Psi Association are not compelling. I came to this appraisal after visiting several séances with different commercial mediums promoted at this location. In particular, I found that I was able to free myself very easily when I re-enacted as closely as possible the binding methods I once applied to one of these mediums, Bill Meadows. This happened as follows: At a commercial séance held with this claimed medium on October 10, 2010, I was invited to tie cable binders around Meadows' wrists to secure his arms on the arms of the chair. On both arms, these cable ties were led through an envelope of jeans material to render them more comfortable on Meadows' skin. These envelopes were closed with a Velcro fastener, and the ends of the cable ties stuck out on both ends of these envelopes. I fastened these cable ties around Meadows' wrists until he told me to stop because they would cut into his wrists if I tightened them further.[2] Indeed, the cable ties in their jeans envelopes seemed to rest tightly around his wrists.

When the séance was over, I cut the cable ties to free Meadows' wrists with pliers and took the cut cable ties home with me. Using a photograph of the jeans envelopes that had been used, I constructed a jeans envelope of the same size and appearance, including its Velcro fastener. Next, Werthmüller kindly informed me about the precise dimensions (height and width) of the armrest of the chair used at the séance, and I built a wooden beam of the same height and width. Finally, using the cut cable tie from the séance with Meadows as a template, I tightened the same type of cable tie around my right wrist and the wooden beam to exactly the same position as in Meadows' séance.

Interestingly, my wrist didn't feel any pressure from the cable tie in its jeans envelope, and I was able to slip my hand in and out of this supposed bond very easily. Of course, one may argue that all this is irrelevant because Meadows' wrist and hand might have been larger, and that my jeans envelope was not a perfect replica of Meadows' envelope. However, I am sure that Meadows' wrists and hands are not considerably larger than mine, if at all, and that also the possible differences in the dimensions of the jeans envelope are irrelevant. It was so easy to slip out of this bond that it would have required enormous and clearly visible differences regarding the dimensions of our hands and jeans envelopes to prevent Meadows from removing his hand from the ties, and such differences definitively did not exist.

In any case, all séance-room phenomena produced by Meadows could have been easily produced by normal means, given if he had slipped out of his bonds. Indeed, it has long been known that supposed mediums who were fastened to their chairs with similar bonds can free themselves easily during sittings. A somewhat famous example concerns a young and likable lady from Brazil, Dona Iris, who gave a sitting for 800 guests in a German town in 1965. Hans Bender, the then head of the Institute for Frontier Areas of Psychology and Mental Health in Freiburg, was very skeptical of public mediums, and he thus secretly observed the séance of Dona Iris with infrared spectacles he brought with him. Bender quickly discovered that she had slipped out of her bonds and impersonated the "spirits" herself (Anonymous 1967, Geisler 1965). A prominent recent example of how a fraudulent medium was unmasked concerns Gary Mannion. He was long surrounded by suspicions—especially after a former circle leader and very close friend of his, Michael Mayo, noted several times in dim light that Mannion impersonated the "spirits" himself. At a later séance in November 2014, Mayo secretly examined the chair to which Mannion was bound when sitters at the other end of the room claimed they were touched by a "spirit" in the dark—and indeed the chair on which the entranced Mannion was supposed to rest motionless was empty (Anonymous 2016a, Anonymous 2016b). As a result of even more suspicions (Whitham 2016), Mannion was secretly filmed with an infrared camera during a séance on May 1, 2016. This very instructive and recommendable documentation of what really happens during séances with Mannion is available on the Internet (Anonymous no date).

In general, it appears to be easy to dupe guest sitters who fasten the ties around the arms of mediums. Often, they never secured mediums before, and they are not trained to examine the critical details (I also speak for myself when I was unexpectedly appointed to secure Bill Meadows).

As a professional magician informed me, one possibility to manipulate the binding simply requires lifting one's wrist a little from the arms of the chair while pretending that the wrists rest on it. A sweater or a shirt with long sleeves is obviously useful to disguise the precise position of one's wrists. Thereafter, there should be enough extra space to free one's hands, and the rest is easy to accomplish (e.g., removing and replacing gags and other ties, rearranging clothes, and moving all kinds of objects in the dark). I suppose that this or a similar technique can also be applied when a supposed medium is tied with duct tape, as was the case when Kruse recorded the voices of Caylor's "spirits." Hence, the duct tape would not even have to be unfastened audibly to free one's hands, as Kruse (2018a) believes. Indeed, at the séances visited by Kruse, Caylor wore a sweater with long sleeves, and a photograph of Caylor after one of these séances at the Basel Psi Association shows on close inspection that the duct tape on Caylor's left arm looks rather loose (Basler Psi Verein 2016). That Caylor's hands are indeed free during some of his séances is proven by the fact that his hands are sometimes tied to different parts of the chair at the closing of the sitting compared with the beginning. These miraculous occurrences are attributed to "spirit" by his followers, who also seem to consider it impossible that he uses his freed hands to produce séance phenomena.

As for Mannion, he simply let the cable ties be fastened around the thickest parts of his forearms and calves, which obviously renders it very easy to slip out of them (Whitham 2016). For an example of how this might have looked, see the photograph of "medium" Mychael Shane taken before a séance at the Basel Psi Association (Kruse 2017e, 2017f). Arm controls of this kind are completely useless and nothing but eyewash. Of course, Shane's arm should be properly be secured. Otherwise, for the sake of honesty, the controls should better be relinquished and Shane's arms left free.[3]

Similarly, it is not as trivial as it might seem to apply proper leg controls. They only make sense if the front legs of the chair a medium sits on are connected with other legs via stable horizontal structures close to the floor. Otherwise, even if cable ties are tightly fastened around the thinnest part of a medium's lower legs, one simply needs to remove the *chair's legs* from the binds by lifting them upward—not the medium's legs!—and the feet are free. This simple trick can be prevented by fastening the cable ties sufficiently tightly around the mediums legs and *above* the potential horizontal connecting structures between the chair's legs. If, on the other hand, such structures are located at a height in which the cable ties can only be fastened around the thicker parts of a medium's lower legs, then the legs can be removed, and the ties are useless again! These very basic aspects need to be taken into account when a medium is supposed to be secured properly. Moreover, using a chair that creaks audibly when people who sit on it move is recommendable (Nahm 2014), especially for "physical mediums," who like Caylor are said to rest entranced and motionless on their chair during sittings.

Commercial séances with "physical mediums" at the Basel Psi Association are quite lucrative. Séances such as those visited by Kruse are typically attended by about 20 people, and at present the participation fee is 180 Swiss francs per person (www.bpv.ch). In my opinion, participants of such expensive public sittings should be able to feel certain that the ties to secure supposed mediums are applied in a way that definitively excludes the possibility that they can slip their hands and feet out of them. Regarding Kai Mügge (Nahm 2014, 2016, Braude 2014, 2016), I recommended several suggestions to improve his control, including the use of a creaking chair (Nahm 2014), and I suggested them (and further possibilities) to Werthmüller in email correspondence between June and November 2014 as well. Yet, Werthmüller didn't think that more stringent controls were necessary because he considered Mügge and Caylor his friends and genuine mediums (email communication to the author on November 26, 2014). Hence, even simple and non-invasive ways to improve the controls are not implemented at the Basel Psi Association. In the light of the above-said and of numerous other aspects not mentioned, it is hardly surprising that many visitors to such commercial sittings hold the opinion that the measures to control supposed mediums at the Basel Psi Association are insufficient.[4]

## Comment on Suspicious Aspects
## of Warren Caylor's Alleged Mediumship

In the context of the topics described in the section above, it might be of interest that accusations of fraud have been repeatedly advanced against Caylor (a critical overview about suspicions surrounding his alleged mediumship is provided at http://www.spiritualismlink.com/t91-warren-caylor; for Caylor's website see http://www.warrencaylor.co.uk/). As it seems, all physical phenomena during Caylor's séances could easily be staged—provided he would be able to free himself from the bonds. Caylor is a friend of Meadows, and both have given several private séances together. On these occasions, "spirits" of both mediums materialize, communicate with each other and also with the guest sitters in the dark (listen to such an event at https://app.box.com/s/n4nqgy3vc4zuogz499l4). This is useful to know because if the "spirits" of one of these mediums are fake, then those of the other medium must be fake as well. At Caylor's séances, voices of many different fully materialized "spirits" can be heard. They include famous celebrities of the past such as "Louis Armstrong" and "Winston Churchill" (recordings of these and many other "spirits" are available at for example https://wcaylor.wordpress.com/tag/caylor/). Also "Michael Jackson" materialized at a séance Kruse visited.[5]

Personally, I never sat with Caylor. Still, I consider it worthwhile to summarize the most important criticisms of his alleged mediumship that are publicly known and available, because most readers will most likely not be familiar with them. Among the many persons who claimed that Caylor freed himself from his bonds on the chair and produced séance phenomena himself are two guest sitters who stated that they clearly recognized Caylor standing in the middle of the séance room, waving glow-sticks in the dark, when the lights of a passing car shone through cracks in window panels (Bland 2008). Similarly, Caylor's former friend and year-long public supporter, Frank Brown, claimed eventually that he and several other sitters recognized Caylor moving around a séance room in comparably bright light. Caylor impersonated one of his "spirits," an American Indian named "Yellow Feather," and spoke with the latter's typical voice. All present sitters except for one clearly recognized Caylor, and immediately thereafter his formerly stable home circle was disbanded by its disillusioned members (Anonymous 2009).

Further instructive reports of sittings with Caylor were provided by Riley Heagerty (2009), who described, among numerous other suspicious aspects of Caylor's mediumship, how the latter was recognized two times walking around in the séance room while pretending to be "Yellow Feather." The first time, "Yellow Feather" manipulated the CD player, and erroneously hit the bass woofer switch which turned on a red light of the CD player. It illuminated Caylor in all his clothes and typical haircut. Caylor then shuffled with small thumping steps back into the cabinet, and Heagerty stated that he had heard the same thumping sounds during many previous séances when Caylor's "spirits" had been active. On the second occasion, "Yellow Feather" left the cabinet carrying an apparently rather bright spirit light. With the exception of one guest who didn't wear glasses that night, all other sitters present clearly recognized Caylor, who spoke in the voice of "Yellow Feather," and shuffled around in front of them in the clothes he wore when he entered the séance room and the cabinet.

It appears particularly suspicious that around that time, Caylor insisted on being fastened with cable ties around his wrists, but with ropes and a knot around his legs (Heagerty 2009). When Heagerty and other sitters were once allowed to examine the empty chair in very dim light after Caylor's body had allegedly been dematerialized (but still might be hidden in a dark corner of the room), the empty cable ties were still on the chair's arms, but the ropes for the legs were nowhere to be found. After another séance witnessed by Brown, the knots securing Caylor's legs differed considerably from the sophisticated knots that were installed before the séance, indicating that Caylor might have loosened the original knots during the séance and replaced them later with more ordinary knots in the dark (Anonymous 2009). Heagerty (2009) also noted with concern that "Yellow Feather's" hands always felt just like Caylor's own somewhat characteristic hands when they touched them.

More recently, Caylor was allegedly caught cheating in early November 2014 in the course of giving séances at the Wallacia Development Center in Australia, once more having seemingly freed himself from the bonds on the chair and enacting alleged spirits. He was sent straight back home to England. Thereafter, he publicly announced on December 4, 2014, on the forum *Physical Mediumship For You*, an Internet forum for physical mediumship (http://physicalmediumship4u.ning.com), that he will start to develop holding séances in candlelight or red light from start to finish to prove that his phenomena are genuine. Moreover, Caylor claimed that in the future the use of cameras would be welcome during his séances (this episode can be followed at http://www.spiritualismlink.com/t91-warren-caylor). In fact, somebody who participated in a sitting with Caylor in 2017 asked him about the possibilities of filming his séances. This sitter informed me that Caylor affirmed that his séances could be filmed at any time with his consent, he only didn't like to be filmed secretly.

Until the present, however, all this is still not the case, thus representing a typical example of what I called "promissory mediumship." Promissory mediums continuously try to keep the interest in their mediumship alive by advancing promises regarding future developments of phenomena and control methods that are finally never kept—or are kept in only such a form that they always remain unsatisfying (Nahm 2015). Similarly, Caylor has shown a conspicuous aversion to scientifically motivated attempts to control his body during séances, although he also stated in public that he is eagerly willing to be tested under any condition deemed necessary by scientists (e.g., Anonymous 2008). Admittedly, Caylor claims that he once was secured properly at an event at Castle Vale in 2008. It seems, however, that no experienced scientists were present on this occasion, and the available descriptions of the applied controls and the present sitters are too general and too credulous to draw reliable conclusions from them (Jon 2008). Nevertheless, his former friend and defender Brown was present on that occasion, and, as mentioned before, he withdrew his support of Caylor not long after (Anonymous 2009).

Yet, given that Caylor had allowed the installation of multiple microphones in the séance room that continuously recorded what went on, and which rendered the retrospective visualization of the location and the movements of his "spirit voices" possible, and given that Caylor was "very excited to literally see (in the 3D representations) what is going on in his séances" (Kruse 2018a:55), and that he announced permission to hold complete séances in dim light and/or to let them be filmed, one should reasonably expect that Caylor will now finally welcome the use of thermal imaging. After all, there is practically no difference between this documentation technique and that employed by Kruse: Microphones passively record audio signals that allow for a retrospective visualization of what goes on, and thermal cameras passively record temperature signals (precisely speaking, electromagnetic radiation) that allow for a retrospective visualization of what went on. Kruse owns the necessary equipment for thermal imaging, and he even recorded a séance of Mychael Shane at the Basel Psi Association (yet, because no phenomena occurred in the space between the circle of sitters, he didn't record anything of significance regarding the question if such phenomena might have been genuine or not; see Kruse 2017e, 2017f). Therefore, I hope Kruse will insist that Caylor now let him use thermal imaging during typical séances as well. In this manner, Caylor and his followers, but also his critics, could see even better "what is going on in his séances."

### Comment on Eckhard Kruse's Analyses
### of Warren Caylor's "Spirit Voices"

Finally, I'll add a comment on Kruse's formant analyses of Caylor's "spirit voices." Kruse analyzed the voices of several "spirits" of Caylor, including that of the notorious "Yellow Feather," who, as described above, was reportedly identified as Caylor himself on several occasions, even by former friends and circle members of his. However, Kruse put the most weight on the characteristics of the voice of a little "spirit boy" called "Tommy." It seems difficult to tell who "Tommy" is. Several years ago, "Tommy" stated that he never lived on the earth plane (Heagerty 2009). More recently, Caylor informed a sitter who visited one of his séances that "Tommy" spoke only Russian when he first appeared in his séances, and then had to learn English. This must have worked well, because "Tommy" speaks fluent English without the slightest Russian accent. Even more curiously, "Tommy's" voice sounds utterly disguised, like a very coarse whispering of a grownup man, and not at all like a child's voice (for recordings of "Tommy" and several other "spirits," see for example https://wcaylor.wordpress.com/tag/

caylor/). However, Kruse stressed that according to his voice analyses, and to formant characteristics of children as presented by Huber et al. (1999), the formants of "Tommy's" voice nevertheless conform to those of a child. This is the core of Kruse's argument, because he believes that it is impossible for grownup men to shift their formants F3 or F4 into the range of 8–10-year-old children. For example, the frequency of F4 of Caylor's normal voice differs by about 1 kHz from "Tommy's" F4 frequency.

To demonstrate that even professional comedians who purposefully disguise their voice are unable to match the formant differences that exist between Caylor's and "Tommy's" voices, Kruse compared the formants of Caylor's "spirits" to those of German comedian Marc-Uwe Kling and his "Kangaroo" (e.g., Kruse 2016b, 2017a, 2017b, 2017c, 2018a). But obviously this approach is illegitimate from a scientific perspective. In comparison with Caylor's "spirits," Kling changes his voice only very slightly when he impersonates the "Kangaroo," and thus it is no surprise that the formants of Kling's original voice differ only slightly from those of his "Kangaroo." Evidently, the legitimate control group Kruse should have employed would have consisted of grownup men who tried to imitate Caylor's "spirits" as closely as possible. Then, Kruse should have analyzed the differences in the formant frequencies of their normal and their disguised voices to put them in relation to Caylor's "spirits."

Since I was asked several times by critical participants of Caylor's séances about my opinion on Kruse's investigation, and because Kruse announced on his website that he would perform further comparative voice analyses in addition to the analyses of Kling's voices, and invited interested readers to contact him, I eventually asked him whether he intended to perform further comparative voice analyses using an appropriate control group as described above. However, Kruse brusquely refused. Apparently, he firmly believed that nobody can raise F3 or F4 by 1 kHz. He claimed my suggestion would question the foundations of voice forensics, and would thus constitute a waste of effort he didn't want to deal with (email communication to the author on December 7, 2017).

Somewhat surprised by this reply, I attempted to imitate "Tommy's" voice, and to analyze it myself. I recorded vowel samples and analyzed their formant frequencies with the software *Praat*, the program used by Kruse. These analyses seemed to show that when I imitated "Tommy's" voice, formant F4 can easily be raised by 1 kHz or even higher, compared with my normal voice. In general, it is not difficult to perform the basic formant analyses with *Praat*. Yet, there are a few stepping stones in that, for example, *Praat* doesn't always distinguish properly between F4 and F5 when the frequency of F4 reaches close to 5 kHz. Nevertheless, even when I chose

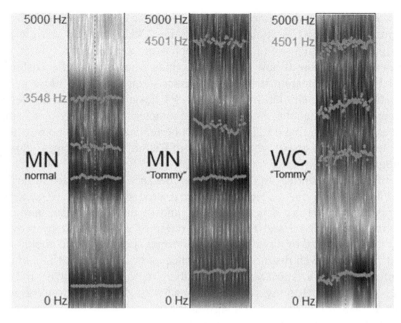

**Figure 1. Graphical display of the formant analyses of samples of the vowel /e/, analyzed with the software Praat.** The four red bars on each subplot represent formants F1–F4, the latter is located at the top. The analyses show that I can lift the frequency of F4 by about 1,000 Hz when I imitate the vowel /e/ as it is spoken by the supposed spirit "Tommy," and that my "Tommy"-F4 is practically identical to that of the real "Tommy" (MN normal = /e/ in Michael Nahm's normal voice; MN "Tommy" = /e/ in Michael Nahm's voice imitating "Tommy"; WC "Tommy" = /e/ drawn from a recording of a séance with Warren Caylor). The three vowel samples have a length of about 0.3–0.5 seconds.

different settings for the formant analyses, the results showed that I was able to raise F4 by about 1 kHz. Still, to be on the safe side, I sent examples of my vowel recordings to a professional linguistic laboratory to let their formants be determined, but without providing further information about the nature of these samples and the reasons for my request. And indeed, the results returned by this laboratory displayed a difference of about 1 kHz in F4 in my vowel samples, and thus confirmed the overall correctness of my formant analyses.

The left part of Figure 1 shows formants F1–F4 of the vowel /e/, spoken with my normal voice. The four formants are indicated by the four red bars. F4 lies at about 3.5 kHz. The middle part of Figure 1 shows F1–F4 for the vowel /e/, but this time I imitated the typical coarse whispering of "Tommy." F4 now lies at about 4.5 kHz, i.e. 1 kHz higher. The right part of Figure 1 shows F1–F4 of a vocal sample from a séance of Caylor's, and gives an impression of how "Tommy" sounds when uttering the sustained vowel /e/. It is apparent that these formant frequencies and those of my imitation are very similar. In both cases, F4 lies at about 4.5 kHz. Hence, it seems likely that the main formant characteristics of Caylor's "spirit voices," including that of "Tommy," can be replicated by anybody who is able to imitate the crucial characteristics of Caylor's "spirit voices" closely enough.

In addition, contrary to Kruse's belief that according to voice forensics, the F3 and F4 of one's voice cannot be altered significantly (e.g., Kruse 2016b, 2016c, 2018a), it is in fact long known that people can modulate the frequencies of F3 and F4 within a range of up to 1 kHz by moving their larynx upward or downward in the normal speaking voice pitch—that means, without even disguising one's voice as drastically as it is required to imitate Caylor's "spirit voices" (Sundberg & Nordström 1976). Indeed, when I imitated "Tommy's" voice, I needed to draw my larynx greatly upward, which must result in a significant lift in formant F4's frequency. Hence, Kruse's reiterating references to voice forensics and the alleged invariance of the frequencies of F3 and F4 in his publications, talks, and interviews are misplaced—and even more so because comparative voice analyses performed in voice forensics are hardly concerned with voices that are disguised in such an extreme manner as required to imitate the voice of, for instance, "Tommy."

The analytical approach I chose, namely using vowel samples of sustained monophthongs to determine their formant frequencies, is a common approach in voice analyses (e.g., Sundberg & Nordström 1976, Huber et al. 1999) because the formants of vowels rather than those of consonants determine the dominant characteristics of one's voice, and because the formants of sustained vowels can be determined most reliably (O'Shaugnessy 2008).[6] My approach lies also at the heart of Kruse's own argumentation which builds on the study performed by Huber et al. (1999), who used recordings of the sustained vowel /ɑ/ to determine typical formant frequencies of children of different ages. Summing up, Kruse's belief that it is unlikely that a single person can (re-)produce the different voices of Caylor's "spirits" including their formant frequencies (e.g., Kruse 2016c, 2018a), and that his analyses "knock the bottom out of the usual fraud hypotheses" (Kruse 2016b:95; my translation), is neither supported by appropriately obtained experimental data nor by the available literature on voice analyses.

## Concluding Comments

To conclude, Kruse only documented and visualized what is trivial and obvious to any participant in Caylor's séances—namely, that different voices move around in the dark space between the sitters. Kruse's approach is, at least in its present form, unsuited for evaluating the possible nature of these voices. An interesting audio-based alternative for testing the authenticity of alleged "spirits" such as "Yellow Feather" and "Tommy" consists in applying linguistic assessments of their speech (Thomason 1989). What is worse, the ongoing dissemination of Kruse's untenable interpretation of his study misleads the public and creates wrong expectations concerning the nature of the documented séance-room phenomena, and it thus does an unfortunate disservice to serious research into alleged phenomena of physical mediumship.

Other technologies such as thermal or infrared imaging are much better suited than Kruse's approach of audio signal processing to assess the origins of voices perceived in dark séance rooms. Because Caylor has repeatedly offered to let his séances be filmed, Kruse should take that opportunity and use his thermal-imaging equipment. Moreover, applying responsible and more stringent controls of alleged mediums is indispensable for future studies of mediumship with a scientific entitlement—especially when suboptimal methodological approaches such as audio signal processing instead of visual documentation techniques are employed. Although Kruse utilizes innovative technical devices and skillfully creates software scripts that generate colorful images, his research takes the second step before the first step, and this second step lacks a reasonable methodological approach and the necessary critical distance (for a more promising example, see Gimeno & Burgo 2017). One hopes Kruse will not have to repeat the words of the disillusioned Frank Brown regarding his earlier public promoting of Caylor's supposed mediumship after being closely engaged with him for more than 5 years: "My realisation came very late and the damage has been done, but I'm now trying to redress that" (Anonymous 2009:4).

## Notes

[1] Mügge has even confessed that he purposefully obtained and employed a magician's LED-device, the effects of which he glorified as "spectacular spirit lights" on his blog. He employed this device during the final "ectoplasm" displays of several séances between 2011 and 2013, including a sitting I attended, and at least two sittings at the Basel Psi

Association. This implies that (at least!) the "ectoplasm" and his trance personality "Hans Bender" were purposefully faked when he used this device. In fact, Mügge's former circle leader Jochen Soederling found such an LED-device in Mügge's travel bag, and he informed me and Stephen Braude about this incident in spring 2014 (this entire episode is described in Braude 2016). Nevertheless, Mügge fervently denies all this. As I was informed via several sources, he spreads (in rather unfriendly terminology) the claim in the "physical mediumship scene" that I blackmailed Jochen to fabricate and publish the stories about the LED-device in Mügge's travel bag and his alleged confession to Jochen, and that Stephen Braude published these fabrications although he knew they were false. Nevertheless, it is easy to prove that this is a simple and transparent (and quite unspiritual) fabrication by Mügge, aimed at demonizing me, Stephen, and our work, to save his hide. Of course, there is a mountain of email correspondence, dating from spring 2014 to 2016, among me, Stephen, and Jochen; and, among others, it also includes Robert Narholz, Hermann Haushahn, Lucius Werthmüller, and Mügge himself. This correspondence establishes beyond the shadow of a doubt that the episode concerning Mügge's confession and the LED device in his travel bag is no fabrication, and occurred precisely as described by Jochen in Braude (2016). In one particularly interesting email to me of April 6, 2014, Mügge even submitted that some phenomena of his public sittings were indeed staged—in contrast to phenomena at "scientific" sittings such as those conducted in Austria in 2013 (compare especially Braude 2014:331). Copies of this and of other relevant emails from this correspondence, but also the photo series described in Nahm (2014) and other unmasking materials, are deposited at the Institute for Frontier Areas of Psychology and Mental Health in Freiburg. They can be viewed upon request. At present, Mügge specializes in presenting extremely suspicious full materializations of disguised "spirits" in dim red light. Measures to control him during such displays are not applied.

[2] Conspicuously, at such public sittings for paying guests, the crucial details of the control methods are always prescribed by the "mediums" themselves.

[3] The spirit guides of Mychael Shane are "Ascended Masters of Shamballah" who include Gautama Buddha, Jesus Christ, and a certain "St. Germain" (Shane no date). According to Warren Caylor, the glass jewelry "apported" in his séances (also termed "activation stones," like the glass jewelry "apported" by Kai Mügge) also originates in "shembala" [sic] and is mediated by apparently the same "ascended masters" such as "St. Germain" (see http://www.spiritualismlink.com/t91p200-warren-caylor).

[4] As described in the main text, the quality of arm controls with cable ties can be tested very easily, playfully, and non-invasively. One just needs to tighten a second cable tie exactly in parallel to the first one around a medium's wrists, and cut and remove it before the séance begins. This tie can then be used as a precise template to re-enact the binding of the medium. After the séance, sitters could test themselves to see if they can slip their hands and feet out of cable ties that are tightened to exactly the same position of the template, and thus, also of the original ties the medium's wrists were secured with. It would also be interesting to examine how much far different sizes of hands would make a crucial difference with regard to freeing them. Yet, it is very important to use ties as a template that were obtained *before* the séance, as there are indications that certain mediums fasten their ties further during séances, most likely toward the end and using their teeth; so that when controllers are asked to thoroughly inspect the ties again before removing them in full light, it would indeed seem impossible to slip one's hand out of them.

[5] Interestingly, full materializations of "Louis Armstrong," "Michael Jackson," and the "spirit" of Winston Churchill also appear in séances of other contemporary commercial "mediums." Still, these "spirits" behave quite differently with the different "mediums." For example, the materialized "Louis Armstrong" of David Thompson loves to play a materialized harmonica. For more information about Thompson, see, e.g., Anonymous (2016c) and several threads at http://www.spiritualismlink.com/f5-physical-mediumship. At a séance with Thompson I attended in 2011 at the Basel Psi Association, even the fake "Hans Bender" of Mügge materialized in the dark and spoke to Mügge, who was present as a guest. At the next séance with Mügge in Basel, the latter's "Hans Bender" affirmed his appearance through Thompson's mediumship. These two "mediums" were friends at that time and mutually praised these remarkable events on the forum *Physical Mediumship 4U*, an Internet forum for physical mediumship (http://physicalmediumship4u.ning.com). However, the "Louis Armstrong" of Chris Howarth doesn't play materialized harmonica at all, but he loves to dance with female sitters in the dark (Anonymous 2016d). Conspicuously, the "Louis Armstrong" of Warren Caylor neither plays the harmonica nor does he invite lady sitters for a dance. In Howarth's séances, "Michael Jackson" also drops in sometimes for a dance with sitters, whereas Caylor's "Michael Jackson" doesn't dance with sitters but faintly sings along to his own songs. It is certainly a pity that these famous "spirits" cannot be seen in the dark, and that sometimes they are also difficult to touch. When some of the supposed dancing partners entered the dance floor in Howarth's séances full of expectation, nobody

was there to dance with. They simply had to sway to the music alone and in darkness until it stopped (Anonymous 2016d).

[6] In fact, formants in spoken words are typically determined in vowel sequences with durations of only milliseconds (Rosenberg, Bimbot, & Parthasarathy 2008; see also Morrison & Assmann 2013).

## References Cited

Anonymous (no date). *What Really Happens When You Sit in the Dark with Gary.* https://web.archive.org/web/20170709133130/http://gary.saves-the-whales.com/

Anonymous (1967). Flüchtiges Psi. *Der Spiegel*, February 20, 109–124.

Anonymous (2008). The séance that failed to materialise. *Psychic News, 3970*:3.

Anonymous (2009). Warren Caylor's accuser speaks exclusively to PN. *Psychic News, 3996*:1, 4.

Anonymous (2016a). 'Physical medium' Gary Mannion caught on video faking phenomena. *Psychic News, 4141*, 6–8.

Anonymous (2016b). 'We don't believe you,' say former circle leaders. *Psychic News, 4143*:38–39.

Anonymous (2016c). Widespread approval for thermal imaging testing. *Psychic News, 4145*:34–35.

Anonymous (2016d). Thermal imaging will be part of a broader assessment of physical mediumship. *Psychic News, 4146*:49–50.

Basler Psi Verein (2016). *Newsletter Physikalische Medialität vom Mai 2016.* http://www.bpv.ch/blog/newsletter-physikalische-medialitaet-vom-mai-2016/

Bland, A. (2008). Spirited away: Meet the psychics with an uncertain future. *Independent*, May 24. http://www.independent.co.uk/extras/sunday-review/features/spirited-away-meet-the-psychics-with-an-uncertain-future-5481164.html

Braude, S. E. (2014). Investigations of the Felix Experimental Group: 2010–2013. *Journal of Scientific Exploration, 28*:285–343.

Braude, S. E. (2016). Follow-up investigation of the Felix Circle. *Journal of Scientific Exploration, 30*:27–55.

Geisler, H. (1965). Die Entlarvung des Mediums Dona Iris. *Die Andere Welt, 16*:1088–1092.

Gimeno, J., & Burgo, D. (2017). Laboratory research on a presumably PK-gifted subject. *Journal of Scientific Exploration, 31*:159–186.

Gulat-Wellenburg, W. von, Klinckowstroem, C. von, & Rosenbusch, H. (1925). Der physikalische Mediumismus. Berlin: Ullstein.

Heagerty, N. R. (2009). Three months with Warren Caylor: Observations and rational deductions. *Psychic News, 4011*:4–5; *4012*:4–5; *4013*:4–5; *4014*:4–5.

Huber, J. E., Stathopoulos, E. T., Curione, G. M., Ash, T. A., & Johnson, K. (1999). Formants of children, women, and men: The effects of vocal intensity variation. *Journal of the Acoustical Society of America, 106*:1532–1542.

Jon (2008). *A home seance with Warren Caylor.* http://physical-medium.blogspot.de/2008/10/home-seance-with-warren-caylor.html

Kruse, E. (2015). Sieben Gründe Physikalische Medialität zu erforschen—selbst wenn man sie für Quatsch hält. *Psi-Info, 34*, 94–97.

Kruse, E. (2016a). Physikalische Medialität: zu unglaublich? *Psi-Info, 35*:104–109.

Kruse, E. (2016b). Hören statt sehen—die Erforschung physikalischer Medialität durch Audiosignalverarbeitung. *Psi-Info, 36*:90–95.

Kruse, E. (2016c). *Gibt es Geister?* https://www.exomagazin.tv/gibt-es-geister

Kruse, E. (2017a). *Forensischer Stimmvergleich für Séance-Stimmen-Phänomene.* http://www.eckhardkruse.net/physmed/voiceanalysis.html

Kruse, E. (2017b). *Forensic Voice Analysis for Mediumistic Phenomena.*
    http://www.eckhardkruse.net/physmed/voiceanalysis.html?en
Kruse, E. (2017c). Geisterstimmen? Die Erforschung der physikalischen Medialität.
    https://www.exomagazin.tv/geisterstimmen
Kruse, E. (2017d). 7 Reasons to research physical mediumship. *Paranormal Review, 82*:24–25.
Kruse, E. (2017e). *Messungen in einer Séance mit Mychael Shane.*
    http://www.eckhardkruse.net/physmed/mychael.html
Kruse, E. (2017f). *Measurements in a séance with Mychael Shane.*
    http://www.eckhardkruse.net/physmed/mychael.html?en
Kruse, E. (2018a). Audio signal processing to investigate alleged paranormal phenomena in
    mediumistic séances. *IEEE Aerospace and Electronic Systems Magazine, 33*:52–56.
Kruse, E. (2018b). Untersuchung Physikalischer Medialität mit Hilfe akustischer Messungen.
    http://www.eckhardkruse.net/physmed/index.html
Kruse, E. (2018c). Using Audio Processing to Investigate Physical Mediumship.
    http://www.eckhardkruse.net/physmed/index.html?en
Maier, J. N. (2016). *Jenseits des Greifbaren: Engel, Geister und Dämonen.* Weinheim: GreatLife.
    Books.
Morrison, G. S., & Assmann, P. F. (Editors) (2013). *Vowel Inherent Spectral Change.* Berlin: Springer.
Nahm, M. (2014). The development and phenomena of a circle for physical mediumship. *Journal
    of Scientific Exploration, 28*:229–283.
Nahm, M. (2015). Promissory mediumship. *Paranormal Review, 74*:15.
Nahm, M. (2016). Further comments about Kai Mügge's alleged mediumship and recent
    developments. *Journal of Scientific Exploration, 30*:56–62.
O'Shaugnessy, D. (2008). Formant Estimation and Tracking. In *Springer Handbook of Speech
    Processing* edited by J. Benesty, M. Mohan Sondhi, & Y. Huang, Berlin: Springer, pp. 213–
    227.
Rosenberg, A. E., Bimbot, F., & Parthasarathy, S. (2008). Overview of Speaker Recognition. In
    *Springer Handbook of Speech Processing* edited by J. Benesty, M. Mohan Sondhi, & Y.
    Huang, Berlin: Springer, pp. 725–741.
Schrenck-Notzing, A. von (1924). Der Betrug des Mediums Ladislaus Laszlo. *Psychische Studien,
    51*:129–160.
Shane, M. (no date). *The Ascended Masters.* http://mychaelshane.com/the-ascended-masters/
Sundberg, J., & Nordström, P.-E. (1976). Raised and lowered larynx—The effect on vowel formant
    frequencies. *STL-Quarterly Progress and Status Report, 17(2–3)*:35–39.
Thomason, S. G. (1989). 'Entities' in the linguistic minefield. *Skeptical Inquirer, 13*:391–396.
Werthmüller, L. (2017). *Paranormales und PSI-Phänomene.*
    https://www.hangar18b.com/paranormales-und-psi-ph%C3%A4nomene/
Whitham, N. (2016). Spirit message confirmed no one would be injured if we filmed 'physical'
    séance. *Psychic News, 4142*:36–37.

*Journal of Scientific Exploration*, Vol. 32, No. 3, pp. 596–603, 2018     0892-3310/18

*HISTORICAL PERSPECTIVE*

# Note on the Intellectual Work of William Stainton Moses

CARLOS S. ALVARADO

Parapsychology Foundation
carlos@theazire.org

Submitted February 13, 2018; Accepted March 14, 2018; Published September 30, 2018

DOI: https://doi.org/10.31275/2018/1300
Creative Commons License CC-BY-NC-ND

**Abstract**—Most discussions about William Stainton Moses have focused on his mediumship. This note is a reminder that, in addition to mediumship, such as the spirit communication recorded in *Spirit Teachings* (1883), he contributed in other ways to the study of psychic phenomena, including studies of direct writing, materializations, and spirit photography. Furthermore, Moses wrote about apparitions of the living and out-of-body experiences, and veridical mediumistic communications, and criticized the writings of others, among them physiologist William B. Carpenter. A consideration of this and other neglected aspects of Moses' work, enlarges our view of his contributions to Nineteenth-Century British Spiritualism and psychical research.

## Introduction

In a previous paper published in the *JSE*, I discussed the fact that some individuals connected to the history of psychical research are sometimes neglected in historical accounts (Alvarado 2012). I also mentioned that some historical figures are only partially remembered for aspects of their work, to the detriment of others. One example I briefly discussed, and which I would like to present more information about here, is William Stainton Moses.

Reverend William Stainton Moses (1839–1892) has generally been discussed as a medium (e.g., Myers 1894–1895, Tymn 2015). One of his best-known publications was his book about spirit communications *Spirit Teachings* (Moses 1883), which, like other writings, appeared under his pseudonym M. A. Oxon. My point in this note is to remind current readers of the somewhat forgotten fact that Moses's work on behalf of Spiritualism, and the study of psychic phenomena in general, consisted of more than his performances as a mental or physical medium.[1] He also contributed to the description and analysis of phenomena, and to the critique of theoretical ideas, some of which I will briefly summarize.

## William Stainton Moses

### *Some Important Publications and Studies of William Stainton Moses*

In addition to holding positions such as Vice-President of the Society for Psychical Research (SPR) and editor of the journal *Light*, Moses wrote about general issues related to Spiritualism. For example, in his *Higher Aspects of Spiritualism* (Moses 1880) had sections about the current state and the future of the movement in England, and about its religious aspects. He wished for better séance conditions that would not look suspicious, and this included "the abolition of all means of secluding the medium, and— though I do not lay so much stress on this—of dark circles also" (pp. 36– 37). Of mediums, he further wrote:

> instead of producing a number of astounding phenomena in the dark, would devote their powers to evolving a very few simple experiments in the light, the whole aspect of the matter would be changed. (Moses 1880:122)

A few years later, after he joined the SPR, Moses's name appeared in the proceedings of the Society as one of the workers on the Literary Committee (e.g., Barrett et al. 1884). But his main intellectual work was published in other forums.

One of these forums was the journal *Human Nature*. Here Moses (1874a, 1874b, 1875c) published a series of articles about physical mediumship.[2] In the first one he focused on general developments and his own conviction in the phenomena. He wrote:

> Most of the phenomena which have been evoked in this country have at different times come under my notice, and have been recorded by me, at the time, as carefully and accurately as has been possible. (Moses 1874a:99)

In other articles in the same series, he wrote about raps and movement of objects, the production of perfumes, and lights and music.

Most of the installments of this article were about spirit photography (Moses 1874b, 1875c). Moses accepted the reality of this phenomenon and stated that he was surprised that he found there was so much evidence on its behalf. He was convinced of spirit agency, but believed that in many cases the spirits involved were of a low moral order. Moses recommended students of the topic not to consider some cases produced by this agency that were suspicious, but instead to focus on clear-cut cases, and on those in which communication (presumably mediumistic) had shown the "reliability" and "truthfulness" of the spirits. However, he cautioned:

> Experience, if it prove that some manifestations look tricky, proves also that
> "second thoughts are best," and that some which appears at first sight most
> suspicious, turn out in the end to be suspicious only in appearance, but real
> and true in fact. (Moses 1875c:195)

In another long paper, Moses (1876–1877) presented discussions and classifications of cases of what he referred to as the "Trans-Corporeal Action of Spirit." This included various cases of out-of-body experiences and of apparitions of the living. Moses wrote that the cases he presented here were scattered and in need of "classification and arrangement" (Moses 1876–1877:241). The cases, he stated, were arranged with a principle in mind.

> The principle is the Transcorporeal Action of Spirit. The fact is the posses-
> sion by man of a spirit which manifests its action in diverse ways irrespec-
> tive of and beyond the power of the body. (Moses 1876–1877:241)

This action of the spirit beyond the confines of the body led Moses to argue the survival of that spirit after death.[3]

In *Psychography: A Treatise on One of the Objective Forms of Psychic or Spiritual Phenomena*, Moses (1878) reviewed the evidence for the phenomenon of direct writing obtained via mediums. He presented examples of cases attested by the senses (vision, hearing), cases presenting writing in languages unknown by the medium, and cases obtained in conditions preventing the previous preparation of writing to fake the phenomenon. In Moses' view, psychography was only one of many phenomena

> which testify to the existence of a soul in man, and to its independent ac-
> tion beyond his physical body; an earnest of its survival and independent
> life when released by death from its earthly prison-house. (Moses 1878:6)

Another book was *Spirit Identity*, in which Moses (1879) studied veridical mediumistic communications. This included personal experiences, and communications recorded by others. The author concluded: "Intelligence is perpetuated after the body is dead" (p. 69), and that the "human spirit after its separation from the body loses none of its individuality" (p. 70).

Moses was also very interested in the phenomenon of materialization. He published in *Light* one of the most detailed studies available of the varieties of features of this phenomenon and tests performed as they were recorded in the literature (Moses 1884–1886). Rather than be concerned with evidential considerations, Moses wrote:

My task is to sift by ordinary methods evidence already published, to classify and arrange it, to indicate its value as a contribution to the study of a very difficult problem, and, whenever I can, to illustrate from my own experience what I am quoting on the published authority of others. (Moses 1884–1886:9)

**William Stainton Moses**

Some of the topics covered by Moses were appearances of inanimate objects (e.g., drapery, flowers), hands, faces, full forms (both shadowy and well-defined), as well as forms that were seen to dematerialize, and that were recognized. Furthermore, mention is made of the weighing of mediums and materialized forms.

Like the previous publications, this one about materializations presented a great number of references that are useful today to locate published discussions about this phenomenon. The cases referred to many mediums such as Florence Cook, William Eglinton, Kate Fox, Mary J. Hollis, Cecil Husk, Francis Ward Monck, Mary Showers, Henry Slade, Charles Williams, and C. E. Wood.

While Moses's articles were rich on descriptions of materializations, he also stated that little was known for certain about the process involved in their production, something he had stated a few years earlier (Moses 1877c). Nonetheless, he believed, as many did in his era, that the phenomenon involved forces processed through the medium's body.

### Other Writings

Various other topics in articles include reports of séances for physical phenomena such as spirit photography (Moses 1875d), voices and materializations (Moses 1877e), cases of spirit identity (Moses 1885a), discussions about the defense of mediums (Moses 1877f), mental healing (Moses 1885b), and psychometry (Moses 1885c). Regarding psychometry, Moses wrote that he sent three writing samples from very different individuals to a psychometrist. The writing samples were dissimilar to each other and were only identified by numbers from one to three. Leaving aside some general statements, Moses said that he

could have easily picked out half a dozen such specific statements in each,
which were at once strikingly true of the writer, and quite inapplicable to
the authors of the other writings sent by me. (Moses 1885c:217)

Although Moses was convinced of the reality of the human spirit, its
survival of death, and of spirit agency as an explanation for phenomena,
he reminded his fellow spiritualists about the resistance others had to these
ideas. As he wrote:

I do not expect much from any attempt to force upon the present generation what it is quite unfit for. A study of the pages of the *Zoist* will shew how
the objective, palpable facts of mesmerism, especially of healing of disease,
advocated with infinite skill and courage, failed to make any permanent effect on a generation that did not want to believe them; that had no place for
them in its mind, no use for them in the economy of its life. So it is with all
these matters. There is a time and a place for them, and it is our business to
have them ready when that conjunction occurs. (Moses 1884a:6)

Moses also admitted his inability to explain phenomena. Focusing on
physical mediumship, he stated:

I do not know . . . how a rap is produced, how luminous bodies are created
in the seance room, how in vacant space is built up before my eyes a body
organised as my own . . . (Moses 1890:156)

But he had no doubt about the existence of these phenomena.

In addition, he criticized many of the critics of Spiritualism. This
included Frederick George Lee and Asa Mahan (Moses 1875e), and Edwin
Ray Lankester (Moses 1876b). Another was the well-known William B.
Carpenter (Moses 1877a, 1877b). Moses did not hold back his contempt
for Carpenter's explanations of spiritualistic phenomena: "Dr. Carpenter
seems either to write in reckless haste without understanding or grasping his
subject, or to answer his opponents without ever reading their arguments"
(Moses 1877a:532).

Moses was also critical of the SPR. For example, he wrote:

The Society for Psychical Research seems to be drifting into the attitude
of the Sadducee, believing neither in angel nor spirit, but only in a sub-
conscious self. That is a pity, but can do no particular harm, for angels and
spirits will still persist in spite of denial. (Moses 1890:157)

In the September 17, 1881, issue of *Light*, Moses took over the
responsibility of writing a column that lasted for several years and that

included summaries, commentaries, and critiques about publications, the phenomena, the issues, the critics, and the social aspects of Spiritualism. Regarding explanations of phenomena, in his first column he commented about narratives of stone-throwing, and theosophical ideas about them:

> I have repeatedly seen phenomena which might far more reasonably be attributed to the gambols of a monkey than to the calculated malice of a fiend. These aimless antics, and many perplexing phenomena of a kindred nature, suggest Puck rather than Satan: a sub-human intelligence devoid of moral consciousness, rather than the serious attempts of a departed human Spirit to demonstrate its continued existence . . . . It is extremely difficult to refer such pranks to departed human Spirits. Nor is it fair to say broadly that all Spiritualists assume this position. The present writer, speaking solely for himself, is free to confess that he has long felt it to be one of the most considerable stumbling-blocks in the way of intelligent acceptance of the creed of Spiritualism that such tricks should be put down to the action of our own departed friends. (Moses 1881:292)

Moses also compiled bibliographies (Moses 1889), and published many book reviews. This included comments about the books of Henry S. Olcott (Moses 1875a), Eugene Crowell (Moses 1876a), Hudson Tuttle (Moses 1877d), Alfred Percy Sinnett (Moses 1881), and Epes Sargent (Moses 1882). In an introduction to a book about mesmerism by William Gregory, Moses related the topic to Spiritualism:

> The Mesmeric Trance, again, accidental or induced; the Ecstatic state, in which the mesmerised subject seems to enjoy communion with the world of spirit, and to live in a state sometimes entirely detached from the world of sense; will readily be seen to have their bearing on such experiences as those of Andrew Jackson Davis, and on the familiar state of Trance into which almost all well-developed psychics are accustomed to pass while utterances purporting to come from an alien spirit are made through their lips; or while their vital forces are being utilised for the production of such phenomena as, for instance, those of Materialisation, or Form-manifestation. (Moses 1884b:v)

## Concluding Remarks

My discussion is certainly not complete, for I have omitted topics such as Moses' work in the organization of spiritualist organizations and his critiques of theosophical ideas. But I believe that the information I have presented shows that Moses had a wide range of interests and made many intellectual contributions to both Spiritualism and psychical research in addition to his work as a medium. Moses in fact reminds us of the multiple roles that some figures played in Nineteenth-Century Spiritualism and psychical research.

Regardless of evidential and theoretical considerations, much of this work, particularly the publications about the "transcorporeal action of the spirit," direct writing, mediumistic communications, and materializations (e.g., Moses 1876–1877, 1878, 1879, 1884–1886), remain valuable reference works to locate Nineteenth-Century publications about those topics.

## Notes

[1] I do not want to play down the contributions of mediumship (and mediums) to the development of psychical research, a topic I have discussed elsewhere (Alvarado 2013).

[2] The first three articles were entitled "Researches in Spiritualism During the Years 1872–73" (1874a), while the rest appeared under the shorter title "Researches in Spiritualism" (1874b, 1875c).

[3] See Moses's (1875b) account of the photographs of his spirit while he was sleeping taken at a distance by controversial spirit photographer Édouard Isidore Buguet.

## References Cited

Alvarado, C. S. (2012). Distortions of the past. *Journal of Scientific Exploration, 26*:611–633.

Alvarado, C. S. (2013). Mediumship and psychical research. In *The Spiritualist Movement: Speaking with the Dead in America and Around the World* (Volume 2) edited by C. Moreman. Santa Barbara, CA: Praeger, pp. 127–144.

Barrett, W. F., Massey, C. C., Moses, W. S., Podmore, F., Gurney, E., & Myers, F. W. H. (1884). Third report of the Literary Committee: A theory of apparitions, Part I. *Proceedings of the Society for Psychical Research, 2*:109–136.

Moses, W. S. [under the pseudonym M. A. Oxon] (1874a). Researches in Spiritualism during the years 1872–73: Together with a discussion of theories advanced in explanation of the phenomena. *Human Nature, 8*: 97–11, 161–176, 193–207.

Moses, W. S. [under the pseudonym M. A. Oxon] (1874b). Researches in Spiritualism. *Human Nature, 8*: 313–326, 337–350, 387–397, 425–430, 473–488, 513–527.

Moses, W. S. [under the pseudonym M. A. Oxon] (1875a). People from the Other World, by H. S. Olcott. *Human Nature, 9*:269–286.

Moses, W. S. [under the pseudonym M. A. Oxon] (1875b). Photographing the spirit of a medium in Paris, while his body was asleep in London. *The Spiritualist*, March 5, 119.

Moses, W. S. [under the pseudonym M. A. Oxon] (1875c). Researches in Spiritualism. *Human Nature, 9*:12–21, 81–86, 97–104, 145–157, 193–213, 334–335.

Moses, W. S. [under the pseudonym M. A. Oxon] (1875d). Some recent experiments in spirit photography with Buguet and Firman. *Human Nature, 9*:157–160.

Moses, W. S. [under the pseudonym M. A. Oxon] (1875e). Spiritualism and some of its recent critics: A reply to certain arguments and objections. *Human Nature, 9*:337–350.

Moses, W. S. [under the pseudonym M. A. Oxon] (1876a). Dr. Crowell on primitive Christianity and modern Spiritualism. *Human Nature, 10*:76–91.

Moses, W. S. [under the pseudonym M. A. Oxon] (1876b). Notes on the present crisis. *Human Nature, 10*:548–567.

Moses, W. S. [under the pseudonym M. A. Oxon] (1876–1877). On the trans-corporeal action of spirit. *Human Nature, 10*:97–125, 145–157; *11*:241–264, 289–312, 337–351, 385–400, 433–441.

Moses, W. S. [under the pseudonym M. A. Oxon] (1877a). Carpenterian criticism: A reply to Dr. W. B. Carpenter. *Human Nature, 11*:529–549.

Moses, W. S. [under the pseudonym M. A. Oxon] (1877b). Dr. Carpenter's theories, and Dr. Carpenter's facts. *Medium and Daybreak*, January 12, 17–18.

Moses, W. S. [under the pseudonym M. A. Oxon] (1877c). Form manifestations. *Spiritualist Newspaper*, May 18, 232–233.

Moses, W. S. [under the pseudonym M. A. Oxon] (1877d). Hudson Tuttle's "Arcana of Spiritualism." *Human Nature, 11*:145–164.

Moses, W. S. [under the pseudonym M. A. Oxon] (1877e). A séance with Mr. Williams, and some reflections thereon. *Medium and Daybreak*, March 2, 137–138.

Moses, W. S. [under the pseudonym M. A. Oxon] (1877f). What have been the results of the defence of mediums? *Medium and Daybreak*, May 4, 276.

Moses, W. S. [under the pseudonym M. A. Oxon] (1878). *Psychography: A Treatise on One of the Objective Forms of Psychic or Spiritual Phenomena*. London: W. H. Harrison.

Moses, W. S. [under the pseudonym M. A. Oxon] (1879). *Spirit Identity*. London: W. H. Harrison.

Moses, W. S. [under the pseudonym M. A. Oxon] (1880). *Higher Aspects of Spiritualism*. London: E. W. Allen.

Moses, W. S. [under the pseudonym M. A. Oxon] (1881). The claims of occultism. *Light, 1*:194.

Moses, W. S. [under the pseudonym M. A. Oxon] (1882). Planchette; or the Despair of Science. *Psychological Review, 4*:124–131.

Moses, W. S. [under the pseudonym M. A. Oxon] (1883). *Spirit Teachings*. London: Psychological Press.

Moses, W. S. [under the pseudonym M. A. Oxon] (1884a). Notes by the way. *Light, 4*:6–7.

Moses, W. S. [under the pseudonym M. A. Oxon] (1884b). Preface to the third edition. In '*Animal Magnetism' or Mesmerism & Its Phenomena* (third revised edition) by W. Gregory, London: Psychological Press, pp. iii–vi.

Moses, W. S. [under the pseudonym M. A. Oxon] (1884–1886). Phases of materialization: A chapter of research in the objective phenomena of spiritualism. *Light, 4*:9–10, 19–20, 31–32, 41–42, 51–52, 61–62, 71–72, 81–82, 91–92, 101–102, 111, 121–122, 131–132, 141–142, 151–152, 161–162, 289–290, 299–300, 309–310, 319–320, 329–330; *5*: 485, 497, 508–509, 525–526, 536–537, 548–549, 560, 580–581, 592, 603–605, 615–616, 627–628; *6*: 8, 19–20, 32–33, 44, 58, 68, 80–82, 92–94, 105, 129–130, 135–136, 166, 188, 195, 211–212, 220, 233–234, 253, 263–264, 273–274, 281–282.

Moses, W. S. [under the pseudonym M. A. Oxon] (1885a). A case of spirit identity. *Light, 5*:547–548.

Moses, W. S. [under the pseudonym M. A. Oxon] (1885b). Mental healing in Boston, U.S.A. *Light, 5*:267–268.

Moses, W. S. [under the pseudonym M. A. Oxon] (1885c). Psychometry. *Light, 5*: 217–218.

Moses, W. S. [under the pseudonym M. A. Oxon] (1889). Bibliography of spiritualism. *Light, 9*:84.

Moses, W. S. [under the pseudonym M. A. Oxon] (1890). Personal experiences of Spiritualism, with some deductions therefrom. *Light, 10*:155–158.

Myers, F. W. H. (1894–1895). The experiences of W. Stainton Moses. *Proceedings of the Society for Psychical Research, 9*:245–353; *11*:24–113.

Tymn, M. (2015). William Stainton Moses. *Psi Encyclopedia*. https://psi-encyclopedia.spr.ac.uk/articles/william-stainton-moses

*Journal of Scientific Exploration*, Vol. 32, No. 3, pp. 604–609, 2018     0892-3310/18

# Tribute to Guy Lyon Playfair (1935–2018)

DOI: https://doi.org/10.31275/2018/1332
Creative Commons License CC-BY-NC-ND

Guy Lyon Playfair, who has died at age 83, was an independent scholar active in international paranormal research for more than five decades, specializing particularly in physical phenomena and telepathy in twins. In his career he turned up some remarkable evidence for paranormal events, most notably during the famed Enfield Poltergeist outbreak 1977–1979 in North London, which he investigated together with Maurice Grosse on behalf of the Society for Psychical Research. The author of 17 books translated into a dozen languages and numerous papers and articles, he was still active in research up to March 2018 when he was hospitalized with what proved to be a terminal illness. At his funeral held in London on May 4, 2018, tributes acknowledged that his range and depth of knowledge on psi topics was formidable; it is hard to identify an issue on which he could not lecture professionally or did not have something interesting or original to say.

Born in Quetta, India, the son of British Army Major General Ian Playfair and novelist Jocelyn Playfair, he was educated in Gloucestershire in England and studied modern languages at Cambridge University. After compulsory conscripted service in the Royal Air Force, working as a Russian translator in Iraq in 1956–1957, he pursued a career in journalism, which included from 1961 his working in Brazil for *Life* magazine and a string of international newspapers. It was in Brazil that his serious interest in the paranormal begun after successful treatment by a psychic healer.

His work in paranormal investigation commenced from this time and can be divided into three overlapping phases. Between 1972 and 1976 while resident in Brazil, he actively researched paranormal phenomena across the country which (save for a few open-minded anthropologists) along with much of Latin America, was very much largely *terra incognita* for Western parapsychologists. Beginning very much as a skeptic, he observed instances of psychic surgery performed without anesthetics and followed up claims of spirit possession and influence, working closely with the Brazilian Institute for Psychobiophysical Research (IBPP) established by civil engineer Herman Guimaraes Andrade (1913–2003) in 1961.

Residence in country placed him in a perfect position to trace the origins of a Brazilian banknote supposedly teleported to the United States by Uri Geller, a claim made by Andrija Puharich. His initial investigation and verdict was a skeptical one published in *New Scientist* magazine and in full in James Randi's *The Magic of Uri Geller* (1975). But to the consternation of skeptics he came to doubt the cogency of this explanation after extensive experience with physical phenomena with Brazil, leading him to promote the nation as 'the most psychic country on Earth.' Phenomena he researched included voluminous automatic writings by trance mediums, séance levitations and materializations, vicious poltergeist infestations, claims of black magic attacks, and cases of reincarnation.

What he saw personally, and his in-depth re-investigations of cases collected by the IBPP, convinced him of the reality of many extraordinary phenomena which he detailed in his fascinating book *The Flying Cow* (1975) re-issued as *The Hidden Power* (1976). Throughout, his personal accounts of investigations into psychic surgery, spiritism, and poltergeists show a strong awareness of the need to guard against deliberate and unconscious frauds and the logging of any lapses in controls.

But he was left with no doubt about the reality of the physical phenomena, after following up on some 20 Brazilian poltergeist cases, including at a private apartment in Ipiranga where he and his colleague Suzuko Hashizume succeeded separately in capturing unexplained rappings on tape. When subjected to acoustic analysis, his recordings show the anomalous sound signature identified as a hallmark of poltergeist effects by Barrie Colvin (2010). Having joined the Society for Psychical Research in 1973, he argued in his followup book *The Indefinite Boundary* (1976) that many of these effects matched cases recorded by psychical research and parapsychology in the Northern Hemisphere.

The second phase of his research career began with his staggered return to Britain during 1974–1976. Settling in London, close to the offices of the SPR and the College of Psychic Studies, he set off on the international psi trail. Helped immeasurably by his flair for languages (he spoke Russian, Portuguese, Spanish, and French), he traveled to the United States, Western Europe, and Eastern European countries that were then part of the Soviet Bloc, contacting other researchers and testing psychic claimants.

Taking up the opportunity to study Uri Geller first-hand, he reversed his previous skeptical opinion. However, what most convinced him was an incident after completing a series of tests when, on his going into a bathroom, a wet shaving mirror materialized and fell slowly through the air before him. He stated: 'I became convinced of something many had learned before me: Inexplicable things do happen in the presence of Uri Geller' (cited in his book *The Geller Effect* (1986) co-written with the psychic).

This reversal of opinion concerning Geller was all the more powerful since he had undertaken detailed studies of both conjuring techniques and hypnotism and from his willingness to collaborate with professional magicians who were actually serious about testing psychic powers; he later co-authored the book *A Question of Memory* (1983) with the internationally renowned magician David Berglas and was a close friend of veteran SPR skeptic and Magic Circle member Dr. Eric Dingwall (1896–1986).

It was Dingwall who backed his investigation into what made him most famous in the field of poltergeist activity. This represented the third phase of his career which can be fixed to September 1977 when he responded to the appeal by Maurice Grosse at an SPR Conference in September 1977 for help monitoring an active poltergeist case at the home of the Hodgson family in Enfield, North London. Although due to take a holiday, Guy immediately decided to attend, fully expecting to discover either trickery or hysteria. What he witnessed swiftly convinced him the case involved genuine PK effects. Canceling his holiday plans indefinitely, he subsequently spent 180 days and nights with the troubled family between September 5th 1977 and June 1978, including 25 all-night vigils. More than 140 hours of taperecordings were obtained, resulting in initial transcripts running to more than 500 pages (a substantial number of his original recordings have still to be transcribed). Many details of the case that became known as 'the Enfield Poltergeist' were published three years later in his book *This House is Haunted* (1980), selling 98,000 copies (re-issued in 2012 and 2015).

Particularly controversial was the appearance of a poltergeist voice. Although criticized at the time by John Beloff and others, who had only snapshot opinions of the case formed from limited visits to the property, Playfair had a number of his critics later rescind their opinions and admit that he had been right. A special committee assembled by the Society for Psychical Research, the Enfield Poltergeist Investigation Committee (EPIC), re-examined the witnesses and collected its own evidence, later issuing a 194-page report reaching the conclusion that psychokinetic incidents had indeed occurred in the house.

When later asked if skeptics who criticized the case at a distance had ever attempted to examine the original evidence upon which his book was based, Guy Playfair confirmed that in more 35 years none ever had taken the opportunity to do so. He also drew attention to much positive evidence from the case which still has yet to be published. One factor that hindered acceptance at the time among parapsychologists was his proposal that a

**Guy Lyon Playfair**

Latin American spiritist approach derived from French mystical writer Allan Kardec served as a better theoretical model, and that employing mediums might prove successful in quelling disturbances. In this he was almost a lone voice in the Anglo-Saxon world since most in parapsychology embraced the RSPK (recurrent spontaneous psychokinesis) model (except for a paper by Ian Stevenson in 1972). His ideas influenced a number of popular treatments of the topic such as Colin Wilson's *Poltergeist!* (1981). As Playfair stated in *This House Is Haunted,* the mere 'mention of spirits invariably polarises people into either fanatical believers or total sceptics.' Ultimately, he felt neither spirits nor PK provided a satisfactory explanation for poltergeist outbreaks. The Enfield case remains the best-documented poltergeist disturbance on record, and, to this writer, assessing the evidence from a legal perspective, the totality of evidence for paranormal effects reaches the standard of beyond reasonable doubt.

In 1983 Guy Playfair was one of a dozen parapsychologists and psychical researchers who offered evidence to the Italian court trying nanny Carole Compton on accusations of arson endangering life (see *Superstition:The True Story of the Nanny They Called A Witch* (1990) by Carole Compton and Gerald Cole).

As well as investigating other cases of poltergeists and hauntings, both independently and for the SPR, he undertook studies in survival research and mediumship and joined the experiments in inducing PK effects in séance conditions developed by Kenneth Batcheldor in the period 1985–1993. A later interest was telepathy and shared sentience between twins, and he went on to publish the book *Telepathy: Twin Connection* in 1999; identifying the biological connections between twins as particularly conducive to psi. He also wrote on meaningful coincidences, xenoglossy, and reincarnation, and despite declining health in his last decade turned to translating a large number of original texts on psi topics from Spanish and Portuguese sources and publishing studies of Brazilian mediumship (for example *Chico Xavier: Medium of the Century,* 2010).

That his work did not gain wider recognition can be ascribed partly to structural problems within psychical research; too many academic researchers are reluctant to engage in fieldwork, leaving the responsibility largely to amateurs with divergent approaches. Guy Playfair successfully spanned the chasm between the two camps, belatedly acknowledged with his appointment as a vice-president of the SPR. He did not hesitate to answer critics on their own level, eventually leading him to resign from penning the column 'Mediawatch' which appeared in the SPR's *Paranormal Review* where he assessed mentions of psi in the press and broadcast media; some of his critiques and lampoons were considered excessive. Secondly, in speaking to both the media and lay audiences, he was prepared to robustly criticize and openly express opinions which others often declined to do, at least in public; his independence meant he did was not restrained by academic pressures. Thirdly, and more controversially, in an opinion shared with Maurice Grosse, he stated that many academic critics of field investigation were actually shy of addressing positive evidence for psi and that such reluctance might have a psychological basis (see Grosse & Playfair 1988). A similar view was expressed by British psychical researcher G. N. M. Tyrell (1952).

It is a challenge that remains, along with a large archive he accumulated, which will, in due course, be made available for those interested.

Away from psychic topics, Guy Playfair had a serious interest in music and played the harpsichord and the trombone and enjoyed real ale and beer-brewing.

With regard to his own demise, during his last illness which became apparent in March 2018, Guy seemed quite relaxed, almost nonchalant. "I've put it to the back of my mind," he told me, sitting on his hospital bed after having learned it was a terminal condition, adding "There's a positive side to everything." Clearly, he had no doubts as to survival in some form, and he seemed certain it would prove yet another fascinating experience.

—**ALAN MURDIE**
camghost@hotmail.com

# References Cited

Colvin, B. (2010). The acoustic properties of unexplained rapping sounds. *Journal of the Society for Psychical Research,74*(April):65–93.

Compton, C., & Cole, G. (1990). *Superstition: The True Story of the Nanny They Called a Witch.* Ebury Press.

Grosse, M., & Playfair, G. (1988). Enfield revisited: The evaporation of positive evidence. *Journal of the Society for Psychical Resea*rch, *55*(813):207–218.

Playfair, G. L. (1975).  Chapter. In *The Magic of Yuri Geller* by J. Randi. Ballantine Books.

Playfair, G. L. (1975). *The Flying Cow: Exploring the Psychic World of Brazil.* Souvenir Press.

Playfair, G. L. (1976). *The Hidden Power.* Panther Books.

Playfair, G. L. (1976). *The Indefinite Boundary.* Panther Books.

Playfair, G. L. (1980). *This House Is Haunted: The Investigation of the Enfield Poltergeist.* Stein and Day.

Playfair, G. L. (1999). *Telepathy: Twin Connection.* Vega.

Playfair, G. L. (2010). *Chico Xavier: Medium of the Century.* Roundtable Publishing.

Playfair, G. L., & Berglas, D. (1983). *A Question of Memory.* London: Johnathan Carroll.

Playfair, G. L., & Geller, U. (1986). *The Geller Effect.* Jonathan Cape, Hunter Publishing.

Stevenson, I. (1972). Are poltergeists living or are they dead? *Journal of the American Society for Psychical Research, 66*:233–252.

Tyrell, G. N. M. (1952). Homo Faber: A study of man's mental evolution. *Philosophy, 27*(103):368–369.

Wilson, C. (1981). *Poltergeist! A Study in Destructive Haunting.* New English Library.

Journal of Scientific Exploration, Vol. 32, No. 3, pp. 610–611, 2018    0892-3310/18

BOOK REVIEW

**Wings of Ecstasy: Domenico Bernini's Vita of St. Joseph of Copertino (1722)** by Michael Grosso, translated & edited by Cynthia Clough. CreateSpace, 2017. x + 278pp. $10.83 (paperback). ISBN 978-1973739906.

DOI: https://doi.org/10.31275/2018/1315
Creative Commons License CC-BY-NC-ND

This self-published volume is a valuable and natural successor to Grosso's earlier *The Man Who Could Fly: St. Joseph of Copertino and the Mystery of Levitation*, which I reviewed very favorably in *JSE* 30-2 (2016): 275–278. In the earlier work, Grosso presented the amazing essentials of the career of the Flying Friar, including some detailed descriptions from eyewitnesses extracted from contemporary sources (including Bernini). In this book, Grosso performs the additional valuable service of providing an abridged translation of the most important contemporary biography of Joseph, a book brimming with compelling detailed eyewitness accounts, many taken verbatim during Joseph's protracted inquisition.

Details always matter, but perhaps more so in a case so remote from the present day and so extraordinary with respect to the magnitude of the reported phenomena. I remind the reader that the case of St. Joseph provides the earliest outstanding evidence for human levitation and quite possibly the best from any era. The levitations were observed by thousands of people, often near at hand, in flight (not simply at his destination) and in daylight. Moreover, the reports often converge on fascinating and unexpected striking details—e.g., that Joseph's clothes would not move during his flights, or that he would not extinguish candles as he flew among them.

Moreover, Joseph reportedly caused many dramatic healings (again, richly detailed in Bernini's *Vita*), and his apparent feats of ESP and bilocation are likewise astounding and difficult to dismiss. So this volume takes us more deeply into the life and character of Joseph and regales us with a great deal more material about the phenomena themselves. In my view, this book is indispensable for students of macro-PK and spontaneous psi generally, and (needless to say) especially so for those who can't read Bernini in Italian.

The abridged *Vita* is followed by a mostly outstanding commentary, developing further some topics covered in Grosso's earlier book, and focusing on the strength of evidence and the historical and religious context in which the evidence must be viewed. However, I wish Grosso hadn't succumbed to the facile temptation to proclaim Joseph's levitations to be a clear challenge to reductive psychophysical theories. Levitation is no more challenging to traditional physicalism than are memory, volition, and many other cognitive phenomena (see Braude 1997). Hard-nosed physicalists will simply offer the usual promissory note that if the phenomena are genuine, the physical mechanisms should eventually be discovered. If we're looking for parapsychological evidence that provides

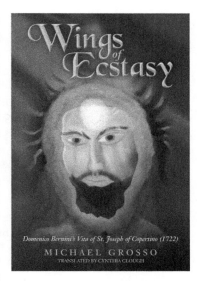

*Domenico Bernini's Vita of St. Joseph of Copertino (1722)*
MICHAEL GROSSO
TRANSLATED BY CYNTHIA CLOUGH

a clear counterexample to reductive psychophysical theories, our only option is good evidence of postmortem survival. Apart from that sort of empirical challenge, the only way to demonstrate the flaws in reductive physicalism is to show (as a number of philosophers have done, in my opinion) that those theories rest on various unintelligible assumptions. But Grosso doesn't take us down that path.

One more niggle about Grosso's commentary. It's marred (but fortunately not subverted) by political pronouncements—mostly, anti-capitalistic rants, which are gratuitous, needlessly contentious, and in some cases, breathtakingly naïve. Consider, for example, this remark, which ignores the central role in terrorism of intolerant religious ideologies: "Terrorism on earth would not exist if the world's wealth were not concentrated in the hands of a tiny percentage of persons" (p. 247).

So, political punditry may not be Grosso's strength. But these lapses in his commentary shouldn't obscure the fact that the book is a major contribution to the parapsychological literature, and that Grosso's essay at the end of it is filled with interesting observations about Joseph's career and the period in which he lived.

—STEPHEN E. BRAUDE

## Reference Cited

Braude, S. E. (1997). *The Limits of Influence: Psychokinesis and the Philosophy of Science, Revised Edition.* Lanham, MD: University Press of America.

*Journal of Scientific Exploration*, Vol. 32, No. 3, pp. 612–617, 2018        0892-3310/18

*BOOK REVIEW*

**I Saw a Light and Came Here: Children's Experiences of Reincarnation** by Erlendur Haraldsson and James G. Matlock. Hove, UK: White Crow Books, 2016. 289 pp. €20.30 (paperback). ISBN: 978-1-910121-92-4.

DOI: https://doi.org/10.31275/2018/1318
Creative Commons License CC-BY-NC-ND

Part 1 of the book comprises the first 166 pages, having been contributed by Erlendur Haraldsson (professor of psychology emeritus at the University of Iceland) (Haraldsson 2017a). The remaining 106 pages, Part 2, are from anthropologist James G. Matlock (Parapsychology Foundation, USA).

Erlendur Haraldsson may be seen as the doyen of European survival and especially reincarnation research. Among many of his books, he has now, at last, come out with one bringing together, as part of his outstanding lifetime accomplishments, his many relevant articles between two book covers (see references on p. 275) (Haraldsson 2017b). He belongs to the pioneers who around the world investigated cases of children spontaneously claiming to remember a previous life.

Part 1 is written in the manner of an empirical field researcher. He lets his cases speak for themselves. Especially the first three examples, but also others farther down, will pose a problem to skeptics to come up with an explanation devoid of reincarnation.

The first case is about the Sri Lankan girl Purnima Ekanayake (p. 3). She made 20 statements about her previous life, 14 of which were correct, 3 could not be tested, and 3 were false. Large birthmarks corresponded to the mode of death in her previous life. Before verification, the previous and the current family were not acquainted with each other; they lived far apart.

The Lebanese boy Nazih in the second example (p. 13) uttered 17 verifiable statements about his previous life, all of which could be shown to be correct. He was asked 15 questions about very private family matters. He could answer them all to the satisfaction of the researcher. The small boy of only 4 years of age recognized a number of members of the previous family whom he could not have known normally. His case is one of the best documented because there were 9 informants who could contribute information about Nazih before contact was made with the previous family. For verification, 4 persons were available.

Example number 3 deals with Duminda Ratnayake (p. 27), a boy from Sri Lanka, who made 8 statements about his previous life which were all correct. Even more remarkable than this is his behavior shown beginning at age 2 to 3 and which suggests a previous life as a Buddhist monk. Professor Haraldsson lists no fewer than 18 behavioural traits relating to this.

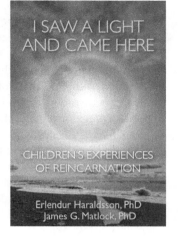

Another boy from Sri Lanka (p. 33), Sandika Tharanga, showed similar clear behavior of a Buddhist monk, beginning from toddlerhood, although growing up in a Catholic family.

Gamaga Ruvan Tharanga Perera's case (p. 36) is similar to the aforementioned. Its peculiarity is to be seen in the fact that written documents had been made before the case was investigated. Apart from this, the boy showed recitative xenoglossy just as Duminda did.

Professor Haraldsson, an Icelander, continues in Chapters 4 and 5 with 5 examples from his home country (beginning at p. 45) that are partly weak and unsolved, demonstrating a clear contrast to the above cited cases.

Because Professor Haraldsson investigated predominantly in Lebanon and Sri Lanka, he presents 3 randomly selected Druze cases in Chapter 6 (p. 57) demonstrating the elements of "recognition," "phobias," "special behaviour of playing," "announcing dreams," and "physiologic reactions."

Professor Haraldsson completes his series of examples in Chapter 7 with three reports from Sri Lanka in which written reports had been made before attempting to solve the case (Dilukshi, p. 73; Pretiba, p. 80; Thusita, p. 84; Chatura, p. 88).

The researcher's case reports are not 1-to-1 copies of the previously published articles but edited texts excluding less important elements of the cases. The different possibilities to explain the material are not discussed in this book except for a short rejection of Professor Keil's notion of "thought-bundles" (p. 164).

Professor Haraldsson contributed to worldwide polls about belief systems and reports on results in Chapter 8 (p. 95). In the US, 25% of the population believes in reincarnation and in Great Britain 24%.

In Chapter 9, Haraldsson acknowledges the achievements of the father of reincarnation research, Professor Ian Stevenson, and his colleagues scattered over the globe (p. 99).

As a psychologist, Professor Haraldsson was interested in, and investigated, the question whether specific properties of the children might explain their peculiar memories of previous lives (Chapter 10:105, Chapter 11:113). For Sri Lanka and Western countries he found characteristic features, but none explaining children's memories of previous lives. It should be mentioned though that some of the children showed signs of posttraumatic stress disorder, which is quite understandable taking into account the high percentage of unnatural deaths of the previous persons.

In Chapter 12, Professor Haraldsson states that children's memories of a previous life prevail longer than hitherto presumed and that no detrimental effects remain from them (p. 124).

The book's title suggests that the book mainly deals with statements and insights about life after death and the question of survival. This is duly covered with respect to children's memories in Chapter 13 (p. 125). However, only a few examples are given, although a statistic from Dr. Tucker, the successor of Professor Ian Stevenson, reveals that 253 children, of a total of 2,242 with memories, report about dwelling in the world beyond, 199 of their burial, and 116 of their conception.

The problem of survival of consciousness is one of Haraldsson's topics. His competence on this matter is exemplified in the following chapters: Chapter 14 (p. 131) deals with deathbed-visions, Chapter 15 (p. 137) is about near-death experiences, Chapter 16 (p. 141) is on spontaneous contacts with the departed, especially apparitions of the dead, and in Chapter 17 (p. 149) we read about contacts with the beyond through mediums. Outstanding examples for the latter field of experience are the medium Hafstein and his contact to the diseased "Runki" who was searching for his missing leg, and the physical medium Indridi Indridason. It is remarkable that among mediumistic contacts, as well as among spontaneous contacts, there are more people who died an unnatural death than are found in the average population.

If reincarnation is real, memories must not exclusively circle around death, but also deal with conception, gestation, and birth. Results of polls around these chapters of life are given in Chapter 18 (p. 157). There are statements of young children saying they had chosen their mothers before coming to earth or had visited the wedding ceremony of their prospective parents.

The second part of the book—beginning with Chapter 20—was written by a profound expert of reincarnation research: Dr. James Matlock. Most of his examples are taken from published articles of the pioneers of reincarnation research. His sequence begins in Chapter 21 (p. 175) with historic cases of children remembering previous lives, some of which were unknown to me. There appear features of "experimental birthmarks" as were later denominated by Stevenson. They are remarkable because no trauma, but mere signs, seem to be sufficient to cause birthmarks.

In the following Chapter 22 (p. 183), Dr. Matlock presents 3 cases of native North American and one of African aborigines. Amongst these many are in-family cases and birthmarks and announcing dreams are prevalent rather than conscious memories.

Most cases worldwide originate from India. That is reason enough to follow in chapter 23 (p. 191) with two examples from India. The first of these is the impressive case of Titu Singh, which was investigated by Prof. Antonia Mills of Canada. Because his date of birth is uncertain two possibilities exist: He might be born 3 month after the death of the previous person (Suresh) or 9 month previously. In the latter case Dr. Matlock offers the explanation that Suresh might have been adamantly insisting on an immediate return, so much that he pushed out the soul of a baby and replaced it with his own soul. Titu's birthmarks would support this model. (Other examples exist suggesting the notion of soul-exchange (Stevenson 1974)).

The second Indian case, Mridula Sharma, was investigated in 1970/1980 by Professor Ian Stevenson together with his Indian colleague Prof. Satwant Pasricha, but was not hitherto published. Dr. Matlock got access to the data, replenished them with new data, and published this case here for the first time. The story is impressive because of its many recognitions and the wonderful relationship which developed between both families concerned.

In chapter 24 (p. 201), the author revives two impressive cases from Brazil. The first being Stevenson's case of Marta Lorenz characterized by the huge amount of 120 written statements of her about a previous life, which regrettably were lost and had to be reconstructed. The previous person Sinhá announced her rebirth which was later corroborated by many features presented by Marta.

The second Brazilian case was investigated by Hernani Andrade, a private researcher. It is about the boy "Kilden" whose mother heard the voice of the (later established) previous person, the Salesian Father Jonathan, when after his car accident he lay in hospital and was in a coma. During her pregnancy with Kilden she had a craving for father Jonathan's favorite food. When old enough, Kilden told his mother that he was once Alexander the priest (Alexander was the second Christian name of Jonathan). Kilden described his accident differently from what his mother had heard on the radio. But his version turned out to be the correct one. Kilden spoke of "his" operation following the accident. What he recalled is similar to the out-of-body experiences that are often part of a near-death experience (NDE). The old argument that the NDEer did not really die and did not come back from death seems to be discredited. More case elements are in the book.

In the next chapter, 25 (p. 209), Dr. Matlock presents a North American case investigated by himself, but being weak compared to the next case from the US. It is about the boy Ryan investigated by Dr. Tucker. Ryan said he was the tap-dancer and movie-actor Marty Martyn. A considerable 55 statements made by the boy turned out to be correct, 15 were wrong and 140 could not be checked.

Dr. Matlock dedicates Chapter 26 (p. 219) to universal and culture-linked patterns of the cases. It is about the percentage of solved to unsolved cases (68% of 2500 cases solved), about the ratio of violent to natural deaths, about the relationships between the families concerned, about the duration of the interim and about the amount of gender-change cases.

Chapter 27 (p. 229) deals with international cases, meaning those in which the previous person is from another country than the current subject. At present only 14 such cases, which have been solved, are known. These are too few to allow for conclusions. But it is interesting to note that in all these cases a motive to cross the border could be found. All deaths in the previous lives were expected, not violent, so that a considered or planned reincarnation was possible.

Xenoglossy, speaking a language never learned, is the topic of chapter 28 (p. 237). Dr. Matlock goes through the few cases of "responsive" and "recitative xenoglossy" we have. He coined the term *passive xenoglossy* meaning cases in which the child learns to speak easily and rapidly because his language is the same as in the previous life. This also implies the counterpart of a child learning only slowly to speak or even reluctantly learns to speak, when the languages of the subject and its previous person are different. Dr. Matlock presents us with one example of "passive xenoglossy" he has personally investigated.

In chapter 29 (p. 245), Dr. Matlock looks for patterns among 9 suicide cases from the literature and one of his own research work. In all cases there existed a social relationship between the current person and members or acquaintances of the previous family, giving rise to the assumption that reincarnation is not governed by mere chance but also is not a process controlled by strict rules. In these cases the median value for the duration of the interim (between death and rebirth) was only 3 months, while it is 16 months in the rest of cases.

In Chapter 30 (p. 255), Dr. Matlock shows cases declared as true cases of the reincarnation type, which may turn up as questionable on closer inspection.

With the last two chapters, 31 and 32 (pp. 261, 267), Dr. Matlock shares with us his thoughts and conclusions on the phenomena found. Questions are raised such as: Why do only a few people remember their previous lives? What is the role of unfinished business? What of us survives death and what reincarnates? Which bodily features and which abilities survive? Can we choose where to reincarnate? Is there a principle like "karma"? Are there soul mates? What about continual development over the course of many lives?

I find it laudable that the book has an index of keywords and that the footnotes are put on each page, not assembled at the end of each chapter, thus avoiding cumbersome searching.

No light without shadow. The first edition of a book is rarely free from mistakes. The appendices A and B announced on page 164 and 220 are just not there. On page 101 my CORT "Rolf" (case of the reincarnation type) is slightly misrepresented (Hassler 2013). Rolf's behavioural memories do not "relate to the opposite sex of the previous personality", as is written, but instead to his homosexuality. This very case is also cited on page 247 as an example for suicide of the previous person although in the original article it is clearly said (on page 22), that it remained unclear whether suicide was intended or not.

In concluding this review I have to say that this book will appeal to believers as well as to open-minded people looking for new horizons. They will find convincing evidence for reincarnation. Most sceptics will not be satisfied with it in spite of its convincing case material; first, because (in my opinion) nothing will ever convince them and second because they cannot quarrel with the authors who do not discuss alternative explanations. If this were to be done carefully it must, in principle, be applied first to each case individually and then to the totality of experiences from all walks of life using arguments of statistics. But this would expand the book to a volume not acceptable to most readers.

—DIETER HASSLER
Independent reincarnation researcher
Germany
dieter.hassler@gmx.de

## References Cited

Haraldsson, E. (2017a). https://notendur.hi.is/erlendur/english/
Haraldsson, E. (2017b). https://notendur.hi.is/~erlendur/english/JournalArticles.htm
http://www.parapsych.org/users/eharaldsson/profile.aspx
Hassler, D. (2013). A new European case of the reincarnation type. *Journal of the Society for Psychical Research, 77.*1(910):19–31.
Stevenson, I. (1974). *Twenty Cases Suggestive of Reincarnation.* Charlottesville: University Press of Virginia.

*Journal of Scientific Exploration,* Vol. 32, No. 3, pp. 618–626, 2018　　0892-3310/18

**UFOs: Reframing the Debate** edited by Robbie Graham. Hove, UK: White Crow Books, 2017. 371 pp. ISBN: B071SK4V3F.

DOI: https://doi.org/10.31275/2018/1321
Creative Commons License CC-BY-NC-ND

## A Cure? Or Another Disease?

UFO studies are stuck in a rut. Mark Rodeghier, Director of CUFOs (The Center for UFO Studies), states that "the field has been stagnating and has run afoul on the rocks of abductions, government coverups/disclosure, and the lack of quality sightings." *UFOs: Reframing the Debate*, an anthology of 14 essays edited by Robbie Graham, is an attempt to respond to that morass by creating a radical new perspective on UFOs. The very first essay, by Canadian science writer Chris Rutkowski, states the dilemma precisely: "UFO belief becomes cult-like when adherents become closed to any interpretation of UFOs as conventional phenomena, and become something closer to religious zealots."

Unfortunately, that insight characterizes most of the rest of the essays in this book. Religious zealots there are, in abundance.

*UFOs: Reframing the Debate* shoots at a high target—the wholesale revisioning of how to approach UFO studies. The contributors intend to alter fundamentally the way we approach the subject. To that end, Graham collects diverse perspectives ranging from a Marxist critique to likening UFO encounters to poltergeist activity. Some of the speculation in these essays is provocative and suggestive. I paused now and again to consider some interesting insights. So the collection does sometimes wing the target—the target being, how to best approach UFO phenomena as a significant field of inquiry, while not locking into an interpretation of UFO phenomena as physical craft—the "nuts-and-bolts" interpretation? But there are no real bulls-eyes. Lapses in logic, imprecise language, and arguments directed mostly at an ETH (extra-terrestrial hypothesis) or nuts-and-bolts bogeyman undermine the intentions of these highly speculative ruminations. Rejection of materialism results in a never-never land of pure thought. I wanted to reach for a balloon full of dried peas to bang them on the heads as Jonathan Swift suggested we do to bring people back to earth. That lack of grounding undermines the seriousness of this undertaking.

The primary unifying theme of the essays is that materialism does not do justice to the complexity of the phenomena, which is true. Accounts of encounters frequently include challenging confusing realities. But that catch-all point of departure does not support the weight of these non-standard speculative forays into what UFOs might be or mean.

The writers explore non-consensual realities—schemes that purport to go beyond what we know to include anomalies that don't compute—but never relate those discussions to the consensual realities in which by definition we live. Anomalies remain mere data if they are not linked to consensual realities in meaningful ways. That reinterpretation does not take place. The anomalies are like floaters in our field of vision, capturing our attention as they flit here and there, but outside our focus.

"Reframing" means a complete restructuring of discussions of UFO phenomena; it should include what came before—the voluminous data that constitute a historical record—and re-contextualize it. These essays don't do that. Despite giving lip service to the reality of material properties of UFOs, those properties are largely ignored or dismissed as irrelevant. That UFOs have mass and energy, represent technologies designed by intelligent beings, and elicited government interest early on because of national security concerns, is a well-documented point of departure for exploring UFOs. But most of these essays reject those facts in order to focus on the impact of UFO experiences on consciousness—altered states, "high strangeness" experiences like the sudden absence of noise during an encounter, telepathic and other paranormal events, quasi-religious experience leading to the formation of cultic communities, and other sociological and philosophical issues. Those are important aspects of some UFO experiences to be sure, but they are not the whole ballgame nor do they support much of the wilder speculations in this collection. To assert that anomalous aspects of some UFO experiences are the ultimate intention of UFOs—an intention without an intender—is as incomplete as ignoring those attributes in favor of an exclusively nuts-and-bolts explanation.

Where we begin a journey often determines where we end up. The editor, Robbie Graham, who produced a highly readable and worthwhile exploration of UFOs in Hollywood films and other media, reports that a point of departure for this volume was the anticipation of "disclosure" in "mainstream UFOlogy" determined in part by the influence of Stephen Greer. I fear that this point of departure derives from discussions of UFO phenomena exclusively on the Internet. Almost all of the contributors to this volume have an Internet presence and publish in that medium and seldom cite serious scholarly mainstream researchers who don't homestead on the Internet. Those scholars are akin to Allen Hynek's "invisible college." They are not household names and do not flit from one UFO conference to another to debate one another (and thereby enhance one another's credibility, like holocaust historians paired with holocaust deniers to bookend talk shows). But they have labored for decades to examine solid data using rigorous scholarly methodologies.

Above all, there is no "mainstream UFOlogy." There are serious researchers, unsupported in the main by research grants, academic respectability, and for the most part, publication in peer-reviewed journals. But the researchers I know are not expecting or distracted by "disclosure" nor do they pay much attention to the work of Stephen Greer. They mine the accumulating data and know the limits of what they know and distinguish it from what they know they don't know. The lines between those two domains blur often in these essays.

An anthology is seldom a coherent whole. An editor's choices matter but so do the decisions of authors to be included in the first place. When the common theme is a negative position without a well-formulated narrative, the result is a lot of firing in different directions. The content in these essays is a smorgasbord of abstractions so almost anything goes. Ideas take off and never land. There is no big picture. So the re-frame is empty.

Some people have approached me over the years with the belief that they had been abducted. When I suggested that they see a mainstream therapist instead of a hypnotherapist in the Mack/Hopkins/Jacobs mold, however, they refused. They wanted a therapist whose conclusions they knew so the therapist's viewpoint could be validated in a circular way. They did not need to be led by the therapist when they led themselves to the therapist.

In a similar vein, those who disagreed with Graham's approach and thematic point of departure would not have wanted in. The omission of contrary points of view is . . . well, a serious omission.

So if UFO research for 70 years is like a bush that needs pruning, what we have here instead is a burning bush, leaving us with ashes. The "ETH or nuts-and-bolts mythology" is replaced with other mythologies but thinner wispier ones with less support. The science-fiction–like epics of wars among aliens, military treaties with aliens, typologies of aliens from tall white to short gray to reptilian, scenarios that fill UFO websites, are not advocated here but they do serve as points of departure for discussing why such narratives exist, why people believe them, what social and cultural organizations result as "real birds" are collected in "digital cages" of online communities.

Such "reframing" frees us from having to pay attention to the mundane origins of our modern preoccupation with UFOs beginning with its inception in the 1940s. At that time, we considered them intelligently designed technologies that challenged our perception and understanding of physics and spacetime and our place in the universe. The USAF Project SIGN concluded reasonably that they were neither Russian nor American technologies, so they did not originate on our planet. The word used to designate that conclusion was "extra-terrestrial," and the concept implied propulsion systems and "flying" and going from point A to point B in space. We said "extraterrestrial" because we did not have ready at hand more exotic categories like multiverse and multi-dimensional reality. We hadn't yet done the math. So we couldn't think of alternatives when we did not have the words to denote them. Einsteinian relativity had been around for half a century but we still thought in Newtonian frames. We thought of moving in space rather than in spacetime.

Math may be the language of physics and cosmology, but most people don't speak it very well. The many books on quantum physics that attempt to translate mathematical concepts into ordinary English do not do the job. There is no Rosetta stone for linking the two languages. We can say the words but we don't really know what they mean. Ordinary language does not give us access to the mysteries of UFO experience either. Inevitably, then, when these essayists try to reframe the debate, the reframing itself is unintelligible because their metaphysical language is arbitrary. The reframing needs to be reframed again. I think of NATO analysts of Russian propaganda who unconsciously absorbed some of the lies they read many times so they needed in turn to be debriefed. Debriefers need to be debriefed, like turtles, all the way down. But that presumes a final debriefer who knows what's true, what's really real, and in this exotic domain we don't have that.

While the USAF Estimate of the Situation used the word extraterrestrial because multi-verses and multiple dimensions were not in our common lexicon, those difficult concepts are now bandied about without regard to their mathematical basis. That obscures the fact that we non-specialists really have no idea what we are talking about when we use the terms. Knowing we don't know the solution to a mystery does not support a leap into the pretense that we do, then using that dubious conclusion as the basis for more leaps like mountain goats springing from crag to crag.

Some writers undermine their sweeping meta-statements with their own meta-reasoning. Greg Bishop, for example, questions the value of eyewitness testimony because the brain co-creates "memories" useful for survival rather than making photographs. "Is it possible," he asks, "that we really make up what we are seeing?" Yet these essays rely on the memories of experiencers to report accurately not only mundane details of encounters but also the elusive content of non-consensual realities. If concerns with memory relate to pilots chasing UFOs or people watching a luminous craft hover over their barns, they certainly apply to reports of out-of-body experiences, telepathic communication, and translation into other domains of spacetime. This is the "post-modernism paradox," which purports to deconstruct the language and meaning of everyone's utterances except one's own.

Rejecting the data that supported the ETH in the first place is the equivalent of throwing out the baby with the bathwater. Throwing out the ability to converse meaningfully by undermining memory and cognition themselves is the equivalent of throwing out bathwater that never held a baby in the first place. It renders all statements meaningless except the statement, "These statements are meaningless."

In other speculation, Bishop advances what he calls "the co-creation hypothesis." The report of a UFO experience, he states, is the result of a creative interaction between the external source of sense data and the subjective response of the human. He says this as if it is something new. Wordsworth referred in his poetry to a common belief of the Romantics that "we half-create and half-perceive" everything, echoing Eighteenth Century philosophers. Scientists do understand that observers influence what is observed. The problem lies not in knowing that but in defining boundaries between observers and what they observe.

Some authors present highly speculative possibilities as if they are facts. Susan Demeter-St. Clair, a professional "research assistant" and psi experimenter, addresses UFOs as a parapsychological event that "may be the key to a greater understanding of the UFO enigma"—which, of course, also means that it may not. But she proceeds as if it does.

Let's accept reports of paranormal effects at face value. Let's forget how unreliable memories are. Let's ignore that the content of such communications—"we wanted you to see us" or "we won't hurt you" might be misunderstood or intentional deception. We have no external point of reference for an alleged communication known only by the "receiver" who reports it. Demeter-St.Clair says that, based on her participation in online discussion groups, she rejects "the wide-held and incorrect assumption that if you do not agree that UFOs are spaceships, then you must not believe that they exist at all—a prevalent view among the ETH crowd," which holds the ETH hypothesis with a fervent blind faith as "firmly entrenched as any religious fundamentalist faith."

An atheist, of course, can be as firmly entrenched as a "true believer," as blindly committed to "no" as a believer is to "yes." Her rejection of those who hold the ETH sounds "as firmly entrenched as any religious fundamentalist faith." That some hold an extreme and untenable viewpoint misses the mark—there is a more reasonable point of view, that UFOs are material objects AND other observations about the phenomena are valuable. That middle ground is often missing in these essays, a gray area that is eclipsed by black-and-white binary distinctions.

Calling the opposition "the ETH crowd" is dismissive and pejorative. Really, are all fundamentalists credulous, blind, unreasoning deplorables? We cannot reject out of hand the notion that intelligently directed technological objects with mass and energy might derive from other civilizations across the hundred million light years of the billions of galaxies in our galactic cluster alone. If other civilizations did evolve, where would they evolve if not on planets in star systems? One can speculate that immaterial skeins of consciousness have evolved in gas giants or nebula, but at that point one can say anything. And if they evolve, how can they evolve if not in material forms? To date, energy and matter seem to be the options. How could they develop societies but through technologies?

Demeter-St. Clair asks of an experiencer, "Were the UFOs she witnessed a psychic manifestation and a cry for attention or help? . . . Is this a case of a poltergeist manifesting itself as a UFO event? . . . was this a UFO event or was it a poltergeist? It appeared to be both." One might as well ask, did the experiencer have a megadose of Vitamin K before the event? Correlation is not causality. The author accepts the account as stated, as an accurate memory, then interprets her question about it as if it turns a speculation into a fact, then turns what it "appeared to be" into what it was. She acknowledges that "UFO witnesses tend to see what they are culturally conditioned to expect" but does not acknowledge the obvious, that so does every human being. "Could the Belgian wave have been a societal cry for help to NATO by a population . . . experiencing very uncertain times?" Hmm, could be. But as essayist Greg Bishop reminds us, "Is it possible that we really make up what we are seeing?"

Or saying. Or saying what someone saw. Or said they saw.

Unsupported conjectures like that are common in these essays. Ryan Sprague tells us that "the majority of witnesses I've spoken with, who've encountered UFOs, have described feeling as if their reality was somehow altered in the moment." This imprecise language does not reference (a) how many witnesses? (b) any documentation? (c) "spoken with"—did you ask if they had that feeling or did they volunteer it? (d) "described feeling" is vague, could you be precise? (e) "somehow altered"—what does that mean? The majority of people watching cable news these days could describe that their reality is "somewhat altered in the moment."

Many essays are written that way, lacking intellectual rigor and discipline, precisely because there are no concrete specifics to document the speculation. Hermann Oberth, the father of German rocketry, said in a speech in 1958 that he had reports from dozens of U.S. Navy and Air Force pilots (and their radar) on which he based his descriptions of UFOs. He reasoned within the domain of consensual realities, admittedly, but so do we all, and he relied on well-documented data from reliable witnesses. He would not, he said, have reported the characteristics of UFOs in granular detail had he not.

Robert Brandstetter describes not consensual realities, the stuff of our current paradigms, but non-consensual realities, and he does it on a grand sweeping scale. "UFO stories arrive out of every culture in the shape of dragons, phoenixes, pearls . . . giant tanks . . . " he tells us, but we are given no evidence that dragons and phoenixes are UFOs, or more precisely, that narratives about them are UFO accounts. He links everything in the sky and makes them all instances of Just One Thing. That One Thing always has traumatic impact, so experiencers struggle to make sense of inexplicable events. "The anomalous experience is a story desperately looking for a way to be told," he claims. But some UFO encounters are not. The data are replete with encounters with anomalous vehicles that are understood as exactly that; see, for example, Tom Tulein's scholarly presentation on the Minot incident (http://minotb52ufo.com/). Pilots often know what they observed and do not show symptoms of trauma.

Reframing discussions of complex domains must be rigorous, well-documented, and inclusive of prior scholarly work. It must demonstrate a familiarity with the historical record. But Joshua Cutchin stretches to try to articulate a point of view "beyond materialism." He writes, "While plenty of cases superficially support the N&B/ETH (nuts and bolts/extraterrestrial hypothesis) view, its materialist foundations are shaken when confronted with the High Strangeness characteristics of a majority of UFO close encounters."

The term "high strangeness" does not become a proper noun simply by using capital letters. Again "plenty of cases" . . . how many? Details? "A majority of UFO close encounters"—Data please? More than fifty percent? What catalogs are used? Discriminating lists or "everybody who calls with a report" à la MUFON? And why does adding aspects of an experience shake the foundations of the initial descriptions when it merely extends and enhances the original narrative?

Are those strange effects a byproduct or an intention of the phenomena? The problem is, we do not know the intention, or meaning, or ultimate implications of UFO phenomena. We know that Nineteenth Century accounts suggest the phenomena have been "here" for a long time, but we know nothing about how "ancient astronauts" or blazing shields in the sky do or do not relate to what we think of as UFO phenomena. Jack Brewer says, "It seems to have been with us a long, long time, whatever 'it' may be," compelling us to ask for documentation once again, and besides, if we have no idea what "it" may be, how do we know "it" has been with us for a long time? Maybe it is a "control mechanism" as Jacques Vallee has written elsewhere (Vallee chose not to submit an essay for this collection), and maybe fairies and gnomes are exactly the same as UFOnauts as Vallee suggests—but maybe not. There is no point of reference for that speculative bridge. Repetition of a belief does not constitute verification. A thousand cut-and-pastes on the Internet do not constitute documentation.

Some speculation is provocative and suggestive, as I said. Some of it made me think. A "nuts-and-bolts" explanation of UFOs is the beginning, not the end of the story. We do need to approach experiencers with empathy to understand the fabric of their experience. But we have been aware of "strangeness" from the beginning, we have noted reports of paranormal effects, and responsible investigators have addressed them as they can. When you don't know what it's all about, you can't get very far. To make progress, we have to explore with a rigorous methodology and historical awareness.

I conclude with the advice of a CIA profiler whose tasks included tracking miscreants through computer networks in order to identify them. She said:

> I look for patterns . . . I try to look with no preconceived notions because the data tells me what I need to know. The task requires intense concentration and constant self-monitoring because there are a thousand puzzle pieces but no box with a picture. So I do not form a pattern too quickly. I observe

myself forming conclusions and ask myself, wait! Is this really true? Or does it only seem true?

When Mike Clelland says in his essay, "Everything is on the table—life, death, sex, dreams, spirituality, psychic visions, genetics, expanded consciousness, mind control, channeling, mysticism, spontaneous healings, out-of-the-body experiences, hybrid children, personal transformation, powerful synchronicity, portals in the backyard, distorted time, telepathy, prophetic visions, trauma, ecstasy, and magic"—everything becomes "nothing" in that grand concatenation of dissimilar things. What seemed like an ice cube on the palm of our hand goes liquid and drips into unintelligibility.

Understanding the matrix of materiality and consciousness in which we are embedded will always be an incomplete project. But paradigms do change over time as anomalies are confronted and factored in, as new hypotheses re-contextualize new data and old and we postulate new theories. To contribute to that enterprise, we need both boldness and humility, and we need to use language with care. We need to keep our loins girded and our lamps lit.

—**RICHARD THIEME**
Thieme Works
rthieme@thiemeworks.com
https://youtu.be/TA8GksT707o

*Journal of Scientific Exploration*, Vol. 32, No. 3, pp. 627–630, 2018     0892-3310/18

*BOOK REVIEW*

**Los límites de la ciencia: Espiritismo, hipnotismo y el estudio de los fenómenos paranormales (1850–1930) [The limits of science: Spiritism, hypnotism and the study of paranormal phenomena (1850–1930)]** edited by Annette Mülberger. Madrid: Consejo Superior de Investigaciones Científicas (CSIC), 2016. 346 pp. ISBN: 978-84-00-10053-7.

DOI: https://doi.org/10.31275/2018/1331
Creative Commons License CC-BY-NC-ND

Inspired by a meeting organized with the support of the Societat Catalana d'Història de la Ciència i la Tècnica [Catalan Society of History of Science and Technology] in April 2013, organized by historians Annette Mülberger and Andrea Graus, this anthology of articles addresses the evolution of Spiritualism and the emergence of pioneers of parapsychology (the old psychical research or métapsychique), mediumship studies, and hypnosis/mesmerism, showing how such practices were introduced and disseminated in Spain and in the European media. Annette Mülberger is a historian of science and Professor of the History of Psychology at the Faculty of Psychology of the Universidad Autónoma de Barcelona and is director of the Center for the History of Science, President of the European Society for the History of Human Sciences, and is currently coordinating a group that investigates the history of science, medicine, and technology of the Nineteenth and Twentieth Centuries. There is also an article about research on Spiritualism in Russia, particularly St. Petersburg, by Dmitri Mendeleev, the creator of the Periodic Table of the Elements.

The book is divided into three parts. After a Prologue under the title "Science, Doctrine, Beliefs and Professionalization," the first part contains three chapters by Mülberger, who describes the origin of the Spiritualist movement in the episode that had as protagonists the Fox sisters, Margaret and Kate, who listened to raps in their house in Hydesville, New York. Since then, the so-called "new Spiritualism" or Spiritism began to expand strongly, both within and outside American borders. In 1857, the work of a schoolteacher and pedagogist Hippolyte Léon Denizard Rivail (1804–1869), would become the leader and encoder of the Spiritualist movement in France. His work, published under the pseudonym of Allan Kardec, consisted of the transcription of messages through automatic writing of the medium Japhet, through whom he wrote the Spiritist "pentateuch," a collection of five writings that became the mandatory reading of the Kardecian spiritism.

In Chapter II, "Spiritism arrives in Spain: The clash with the Catholic church and the First International Congress," Mülberger presents the way in which Spiritualism—as in all Western countries—was introduced among families through "table-turning" through which sitters in a séance could "get in touch" with the spiritual world. By then, important figures of the Spanish cultural world, such as Manuel Sanz and Benito (1860–1911), the Count of Torres Solanot (1840–1902), as well as the Spanish medium Amalia Domingo Soler (1835–1909), contributed to the spread of Spiritualism in all of Spain. The First International Spiritualist Congress held in Barcelona in 1888, and the clash of Spiritualism with the Church, produced a conflict of interest to the point that dozens of books on spiritualism were burned in an Act of Faith [Auto de Fe], among others, in a public bonfire organized by the Catholic Church.

In Chapter III, "The investigation of the paranormal," Mülberger analyzes the different lines of research that arose from a psychic (psi) hypothesis such as the mental activity of the medium but disconnected from discarnate spirits, in order to explore the spiritualist phenomena from a scientific perspective. In the English-speaking literature, these studies were known as psychical research, but in Spain the French literature was well-known as métapsychique, a term coined by the physiologist and Nobel Laureate Charles Richet. In later years the industrialist Jean Meyer and others founded the Institute Métapsychique International in Paris in 1919. Mülberger highlights research studies of the British physicist William Crookes (1823–1913), the psychological theories of F. W. H. Myers (1843–1901) with regard to the "subliminal self," Pierre Janet's (1859–1947) studies about mediumship, and studies of materializations in the laboratory by Baron von Schrenck-Notzing (1862–1929).

In Chapter 4, Andrea Graus presents "Dispelling the spirits: The scientific study of mediumship," in which she develops the idea that spiritualists should create the necessary conditions to "cultivate" mediums, allowing those sensitive people to develop their mediumship through training, and thus contributing to a secular religion. Spiritualism circles were not composed of spiritual leaders but were deeply democratic, so that all people had the possibility to discover their own "channels of communication with the spiritual world, through automatic writing or speakers mediumship, or other exceptional modalities such as painting or drawing" (pp. 112–113). The Spanish Nobel Prize winner Santiago Ramón y Cajal (1852–1934) also showed a keen interest in the Spiritualism movement. Since 1882, hypnotism had a progressive presence in the Spanish medical community, and in Chapter

5 Ángel González de Pablo addresses the process of scientific legitimization of hypnotism in the context of psychology and mainstream medicine. According to him, Spanish doctors such as Juan Giné y Partagás (1836–1902) and Luis Simarro (1851–1921), contributed very influentially in the publications on hypnotism and played a fundamental role in the consolidation of the study of hypnotism in Spain.

In Chapter 6, "The practice of the Metapsychic: A Marquis investigating clairvoyance," historian Mónica Balltondre examines the studies of Joaquin Jose Javier Argamasilla de La Cerda y Bayona, well-known as the Marquis of Santa Cara (1870–1940). He was a Navarrese aristocrat based in Madrid who devoted himself to studying mental experiments on telepathy at home, and published one of the few well-known Spanish books *Un Tanteo en el Misterio* [*An Estimate in Mystery: Experimental Essays on Somnambulist Lucidity*], in the 1920s, based on in his experiences with his son Joaquin, who was well-known as a psychic and visited New York for an interview and a number of experimental sessions with magician and escape artist Harry Houdini. According to Marquis of Santa Cara, Joaquin was better-known for claiming an ability to see through opaque objects. His father began to make demonstrations of his psychic ability for reading paper sheets tucked inside sealed boxes or guessing the hour of clocks (previously handled) also hidden from sight. Among the audience of these shows was the Spanish writer Valle-Inclan, who was a friend of the Marquis of Santa Cara and became convinced that Joaquín's powers were real. In Chapter 7, Nicole Edelman argues that the beliefs of the spiritualists established a distinction between a person sensitive to the influence of the spirits and the sleepwalker who acts under the influence of his own spirit (spiritualistic vs. animistic theories).

In the Epilogue, Mülberger recognizes that the history of this articulation "science and spiritism" requires an interpretation that forces us to situate ourselves in the history of scientific thought of the time in agreement with other Latin American authors who also found similar data in their own historical studies, such as those carried out in Argentina (Gimeno, Corbetta, & Savall 2013, Quereilhac 2016, Parra 1995), Chile (Vicuña 2006), or México (Leyva 2005).

*The Limits of Science* ends with an "Appendix" that includes a definition of the concepts usually named in psychic research, metapsychics, and parapsychology, and a list of References.

—ALEJANDRO PARRA

Instituto de Psicologia Paranormal, Buenos Aires, Argentina
rapp@fibertel.com.ar

## References Cited

Gimeno, J., Corbetta, J. M., & Savall, F. (2013). *Cuando hablan los espiritus: Historias del movimiento kardeciano en la Argentina.* Buenos Aires: Antigua.

Leyva, J. M. (2005). *El ocaso de los espíritus: El espiritismo en México en el siglo XIX.* Mexico: Nexos Sociedad, Ciencia y Literatura.

Quereilhac, S. (2016). *Cuando la ciencia despertaba fantasías: Prensa, literatura y ocultismo en la Argentina de entresiglos.* Buenos Aires: Siglo XXI.

Parra, A. (1995). Research aspects and social situation of the parapsychology in Argentina: Brief history and future possibilities. *Journal of the Society for Psychical Research, 60:*214–228.

Vicuña, M. (2006). *Voces de ultratumba: Historia del espiritismo en Chile.* Santiago, Chile: Taurus.

*Journal of Scientific Exploration*, Vol. 32, No. 3, pp. 631–634, 2018     0892-3310/18

**The Elements of Parapsychology** by K. Ramakrishna Rao. Jefferson: NC: McFarland, 2017. 337 pp. ISBN 978-1476671222.

DOI: https://doi.org/10.31275/2018/1337
Creative Commons License CC-BY-NC-ND

Rao is a major force in parapsychology research, having first published in the 1950s, and this book publication is right up-to-date in 2017. The book is almost 300 pages of text spread out over 9 chapters. I initially found the book title to be confusing because it could be taken to mean something like a primer or an introduction to the essential parts of parapsychology. Whereas in fact the book is more like a selection of discrete areas of parapsychology such as psi missing (the tendency to actively avoid the 'target' of a psi experiment, to the point of statistical deviation away from chance); the experimenter effect (similar to the experimenter effect of conventional psychology, but here the suggestion is that experimenters are using their own psychic powers to influence experimental results); problems with replication in psi experiments; and some other philosophical or epistemological considerations. I'll come back to the question of the title at the end of the review.

There is much to recommend in this book, but there are also problems depending on what one's expectations are in reading the book.

I found myself many times in this book wondering which audience Rao was trying to address. First of all, it is not for the casual reader (i.e. non-scientist), as the text is written in a straightforward 'academic' prose style. There are no diagrams, figures, or graphs, just a few tables to help the 'newbie' to understand the concepts that Rao is espousing. Secondly, many technical terms that would be familiar to the professional parapsychologist, are used without definition, or they are defined much later in the text. One example is that the term 'sheep–goat' (so-called believers vs. nonbelievers in psychic powers) is used early on, but it is defined several chapters later and there is no entry to this term in the index. This leads me to guess that Rao's target reader audience is very firmly in the professional parapsychology camp.

I did find that some of Rao's writings were too brief or condensed and would require considerable unpacking for someone not familiar with the research database to understand effectively. For example, in Chapter 4, on page 119 (Problems of Replication and Application), Rao writes:

> In psi research, the decline of effect sizes is all too common, leading researchers to frequently change their strategies often resulting in reinventing the wheel.

For a variety of reasons I think this is an important sentence but then I know what the 'decline effect' is (I know that it's not quite the same as the 'decline of effect sizes' but it's clear that Rao is referring to the decline effect), and I think I have a vague idea of what Rao is saying when he states that researchers frequently change their [research] strategies, but without him spelling it out I am not too sure. I have to state I'm intrigued when he says that psi researchers are reinventing the wheel, but I am left grasping at straws to understand which wheels he is referring to. I looked for clues later in the book, and if they were there then I missed them. I am reasonably experienced in parapsychological research literature, so I can imagine that the non-parapsychologist reader would probably gloss over this and other points that Rao tends to write in a super-condensed format. The last four or so paragraphs of the same chapter deal with issues around replicability in the Ganzfeld and this is an area that I am more intimately aware of since I met and worked with the main architects of the Ganzfeld, but I would have to say that the reading was so condensed that I was not any the wiser after several attempts at reading it (of course that could just be that I am not as intelligent a reader as I think I am).

I will say that I found Rao's foundation in experimental research rooted very much in the pre-meta-analytic days of parapsychology with his reliance on the '$p$-value,' standing for meaningful significance (instead of this index actually standing for statistical significance). Rao knows about meta-analysis and cites the work done in key areas of parapsychology, but he still does not seem to focus on the 'effect sizes' as being the essential meaningful index of measurement in experimental psychology. In Chapter 6 'The Problem of Psi Missing,' Rao speaks at length about the differential effect (DE) which is the phenomenon that participants tend to score either toward the intention or away from the intention of the experiment because of either different experimental conditions within the experiment, or different types of tasks, or even different types of potential responses. So for instance participants might prefer to respond to ESP (extra-sensory perception) experiments that have positive words as targets, compared with negative words. DE is one of the areas that Rao's own research and publications have focused on. He explains how he and others compiled a review of all the available DE–

ESP experiments and then provided tables to show what percentage of the reviewed experiments achieved statistical significance; so he evaluates the quantitative research literatures by counting a study that reached statistical significance as a datapoint regardless of effect size or number of participants. For instance, on page 182 he writes:

> There are 73 studies in the category of dual target conditions. The subjects in these tests attempted to guess two different sets of targets such as words in two languages. In 46 of them (63 percent) there is differential scoring, i.e., the scores are in the opposite direction for the two conditions, when you expect such a scoring in about one half of them. The probability of 46 studies showing differential scoring in a total of 73 studies is <.05. In 20 of these 46 studies, the difference between the scores obtained under the two conditions is highly significant. The associated p-value with such an outcome is very, very small indeed.

A table is provided on the same page, and it is clear that Rao is evaluating his database in, by today's standard, an outdated way. This part of the text pretty much jumped out of the page and screamed 'do a meta-analysis'! After all, as the original review of the experimental literature was done by Rao and his associates, one would have thought that (perhaps with the help of others) his extensive knowledge of those studies would have enabled him to have a far keener insight into the meta-analytic results, than someone coming in 'cold' to the data.

Parapsychology suffers in my view, by being constrained with an Anglo–Eurocentric Weltanschauung. This is of course a charge that can still be made in psychology today, and the number of active researchers in parapsychology is several magnitudes of order fewer than in psychology. Rao does make mention of parapsychological observations, events, or research done in India and some other places outside of Europe and North America, and personally I thought/hoped that I would read a lot more about that given Rao's education and upbringing in India. For anyone not steeped in cross-cultural psychological research, these mentions require considerably more 'unpacking' than is given. I acknowledge that he has already written on the comparison of a 'Westist' approach versus others in more traditional psychology (Rao 2002). Perhaps Rao's insights into a 'Westist' approach in parapsychology specifically is for another book—I for one would look forward to it eagerly.

Much of the book seems to be rooted in the experiments and research between the 1930s and 1980s. That is not to say that Rao is ignorant of the current research or that there are not current citations in the text, but rather that the preponderance of the citations are firmly established in the 40-year time period mentioned. If one takes the view that this book gives a firm oversight of essential research done in this time period, then this is an excellent current resource to revise one's background understanding of original experimental research done in parapsychology.

To return to the title of Rao's book, I feel that it deserves a different title to help readers figure out what they can get out of the book. I'm not a copywriter so forgive my meagre talents at snappy titles or headlines. I am just trying to propose the spirit of a new title, something along the lines of: "Selected Thorny Issues in Parapsychology," or "Conundrums in Parapsychology: Some Reflections," would have orientated me far better to approach Rao's otherwise comprehensive academic book.

Whom is this book for? I think Rao is writing for an audience of people already in the parapsychology field. It serves as an effective summary of the research findings of the middle and toward the latter part of the Twentieth Century with some updated citations from the Twenty-First Century. I want to stress that this is **not** a criticism but rather a strength of the book. Rao has personal insight by being at 'the coal face' during the majority of this fertile time period in parapsychology, and there is much to commend the quality and insight of this work. Modern parapsychologists might be using key citations from this time period but not with the breadth and depth and critique that Rao provides. All research arenas probably suffer from the lessons of previous generations of researchers or inquirers not being remembered (well), including the nuanced considerations in summarizing the work that might lead to (at the time) future avenues of research. Parapsychology is no different in this regard, and having a book such as Rao's is a welcome addition for the researchers of today to not forget those hard-earned insights of the mid-to-late Twentieth Century.

—**ROBIN TAYLOR**
Oceanik Psi
robin@oceanikpsi.org

## Reference Cited

Rao, K. R. (2002). *Consciousness Studies: Cross-Cultural Perspectives*. Jefferson, NC: McFarland.

*Journal of Scientific Exploration*, Vol. 32, No. 3, pp. 635–637, 2018     0892-3310/18

**Rigor Mortis—How Sloppy Science Creates Worthless Cures, Crushes Hope, and Wastes Billions** by Richard Harris. New York: Basic Books, 2017. 222 pp. $28 (hardcover), $18 (paperback), $12 (ebook). ISBN: 978-1541644144.

DOI: https://doi.org/10.31275/2018/1338
Creative Commons License CC-BY-NC-ND

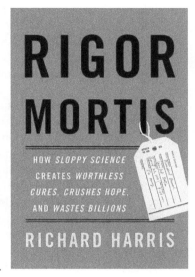

The subtitle of this book is accurately descriptive. The "science" underlying much of modern medical practice, particularly that associated with prescription drugs, is largely incompetent owing largely to faulty statistical approaches; and even when mistakes are pointed out, that does not ensure that practices change.

The fundamental trouble is that the research enterprise has become far too big and far too competitive and is corrupted by commercial and political agendas and influences. One indication of the competition-induced hype and spin is the inflated self-lauding language in articles in medical journals; the frequency of such words as "novel," "unprecedented," and the like increased by factors as great as 150 between 1974 and 2014 (pp. 190–191). The book mentions these fundamental issues but does not sufficiently emphasize them, indeed obscures them by suggesting remedies that do not get at the fundamental trouble: That researchers take more responsibility for good practices, for instance, is just whistling in the wind when the problem is systemic—as the book acknowledges in a few places: "The much harder challenge is changing the culture and the structure of biomedicine so that scientists don't have to choose between doing it right and keeping their labs and careers afloat" (p. 4); "Biomedicine's entire culture is in need of serious repair" (p. 167)—the prevalence of sloppiness, haste, corner-cutting, is amplified by the human penchant to "do what everyone is doing" (p. 187) because under present circumstances that seems effective in bringing visibility and prestige (p. 188).

In reality, official regulation must become competently evidence-based, the influence of Big Pharma needs to be brought under control, and only governmental actions could begin to do the job, including initiatives to make the research enterprise smaller rather than larger.

For people who have already delved into the pervasive errors and commercial deceits prevalent in modern-day medical research and practice (Bauer no date), there is little new here. But the illustrations are well-chosen and powerful, and a few tidbits deserve mention:

- Surprising as it may seem, pre-clinical animal studies have often been less competently done than clinical trials with human beings, and moreover "animal models" of a human disease may not be valid models or analogues (p. 55 ff.). That matters enormously because it is the pre-clinical studies that determine what lines are followed and which are not. One common failing of animal studies had been that researchers used only male mice and not females, for reasons of convenience (p. 41); it was only in 2014 that the National Institutes of Health announced a policy to ensure that sex was recognized to be an important variable (Clayton & Collins 2014).
- Mice are used very widely as convenient animals, but the results can be entirely misleading (Chapter 4, p. 71 ff.). What happens with mice does not always jibe even with what happens in rats! The drive for reproducible results entails that researchers use carefully selected strains of animals; but the results may then be misleading because the tested subjects are too homogeneous to illustrate what happens in the real world.
- A cash incentive showed more apparent improvement than any tested drug with patients suffering from muscular dystrophy (pp. 43–44), more than suggesting that there is something wrong with how improvement is measured.
- Routine lab procedures may have unrecognized effects: cleaning glassware with acid rather than detergent can make a difference (p. 45). The mechanical means used to stir solutions can make a difference (p. 47).
- "Between 18 and 6 percent of all cell experiments use misidentified cell lines"! (p. 96) even after the mistake has been pointed out!! (p. 99). More than 1,000 published articles claiming to be about breast cancer were actually done with skin-cancer cells (p. 102). Needed precautions against using bogus cell lines are far from universally

employed (p. 111).

- The problems with cell lines are mirrored by problems with antibodies. Monoclonal antibodies are supposed to be highly specific but they are not (p. 112 ff.).
- There may be about 12,000 papers based on bogus cell lines, each paper cited an average of 30 times (p. 103). The literature is highly unreliable, in other words; mistakes are not eliminated, they live on. "Years after two of the largest and most expensive medical studies ever undertaken had debunked the claim that vitamin E reduced heart attacks, half of all articles on the subject still cited the original [mistaken] study favorably" (p. 219).
- The searches for genes that influence particular conditions—obesity, schizophrenia, blood pressure, etc., etc.—have been going on for many years. A careful analysis by Ioannidis of publications in this genre identified as reliable only 1.2% of tens of thousands of articles.
- Many journals make it difficult to publish corrections to earlier-published articles (p. 182).
- The flood of publication is exacerbated by official incentives by some nations to gain prestige. Chinese scientists gain cash bonuses for publishing in *Science*, *Nature*, or *Cell*; and the authors multiply their awards by selling co-authorships (p. 178).

—HENRY H. BAUER
Professor Emeritus of Chemistry & Science Studies
Dean Emeritus of Arts & Sciences
Virginia Polytechnic Institute & State University
hhbauer@vt.edu
www.henryhbauer.homestead.com

## References Cited

Bauer, H. H. (no date). *What's Wrong with Present-Day Medicine* (a bibliography, last updated 17 April 2017).
    https://mega.nz/#!oOAhVaxA!BwxcAEUqYP4V5eDDwtPnWGwoJvkUpp5NVaPPD0akNHs
Clayton, J. A., & Collins, F. S. (2014). Policy: NIH to balance sex in cell and animal studies. *Nature, 509*:282–283.

*Journal of Scientific Exploration,* Vol. 32, No. 3, pp. 638–641, 2018          0892-3310/18

**The New Genesis: The Greatest Experiment on Earth** by Wojciech K. Kulczyk. Frimley, UK: The New Genesis Foundation, 2017. 248 pp. $12.95/£10. ISBN 978-1-9999060-0-9.

DOI: https://doi.org/10.31275/2018/1353
Creative Commons License CC-BY-NC-ND

The Universe is a complex place. Science's job is to model how nature works, and to turn our understanding into something useful. Generally, this modeling works on a reductionist principle: Science is at its best when seemingly complicated processes in nature can be understood with simplifying equations and hypotheses. Where physics is concerned, this generally works out pretty well. Likewise with chemistry, notwithstanding some exceptions to the general rules. But when it comes to biology, the complexity involved seems altogether staggering.

In this book, the author grapples with the disparity between the straightforward principles of evolution by natural selection, and the sheer immensity of the task when attempting to successfully apply those accepted tenets to biological processes. Wojciech Kulczyk is not convinced that evolutionary principles are capable of explaining complex forms of life. As a physicist with a Ph.D., he is no stranger to science. His writing is clear and erudite, demonstrating a strong grasp of many of the sciences. Yet, he remains fundamentally puzzled by the evolution of complex life on Earth, and, in particular, by how it could have arisen as a matter of chance.

Kulczyk considers it highly likely that the increasing complexity of life here on Earth was given a helping hand. Actually, many helping hands. He argues for an intelligence behind the design—that life needs an engineer to create the cosmic blueprint that churned out complex life on Earth. He stops short of identifying whether that intelligent designer is a spiritual entity, or a set of interested parties existing in the physical realm. Reading between the lines, he favors the latter.

The author argues that the Earth is such a ridiculously ideal crucible for the emergence of complex life that such a fortuitous location could not have arisen by mere chance. It seems uniquely qualified to host life—a circumstance that the author finds untenable. Instead, a being or beings contrived to set up the stage for life— going to some considerable lengths. This includes the purposeful collision between a comet and the Earth to

provide our world with its late veneer of
life-bearing water. Then, along the way,
'they' intervened on multiple occasions to
create the intermittent bursts of evolutionary
progress noted from the Earth's fossil
records. In a roundabout kind of way, the
author mixes punctuated equilibrium with
alien intervention. Our world is essentially
an experiment in cosmic and biological
engineering, he argues, which we mistakenly
think was either due to random chance or to
God.

There are problems with this reasoning,
I would argue. The Universe is immense
enough for almost any improbable event
to emerge somewhere. Even if this is the
most favorable place in the universe for complex life to emerge, capable of
reflecting upon itself and the circumstances within which it finds itself, then
that's okay. The chances of self-reflective consciousness finding itself in
the best-placed world in the universe for it to emerge isn't infinitely small:
Instead, it's 1 in 1.

Even so, I'm not remotely convinced that life is that precious. The
author discusses *panspermia*. He recognizes that our solar system is
relatively young compared with much of the Cosmos. However, he doesn't
really entertain the notion that life, in some fledgling form at least, can
transfer seamlessly between star systems via interstellar comets, or even
be encountered within nebula way-stations along a star's grand tour of the
galaxy. Why not? Instead of life being only here, why can't it be absolutely
everywhere, spilling out into space and seeding itself on every planet or
comet (usually unsuccessfully)? In which case, the multiple parallel paths
to complexity become essentially infinite in their extent, and the 'chances'
of high-level functionality emerging from mutating systems over long time
periods increases exponentially.

Kulczyk's thesis becomes grittier as he examines the complex nano-
engineering that is cellular biology. The complex functionality of cellular
processes is mind-boggling, and he does an incredibly good job of bringing
this all to life. Photosynthesis is a case in point, with its series of biochemical
processes whose origins seem to defy random mutation. Pitched at the level
of popular science, Kulczyk's descriptions of how cell biology work are
factual, informative, and well-explained. He makes good use of metaphor
and analogies to illustrate his many points, and I was better informed about

modern developments in these sciences as a result. For example, in the following extract he questions how the complex biochemical processes facilitating nitrogen fixation could have emerged by chance, with its protein of about 31,000 atoms that contains molybdenum and iron:

> It is difficult to envisage how evolution could invent such a complex system. how evolution could select such a special metal cluster interacting with dozens of amino acids. Again, DNA codes not only this huge catalyst molecule, but also nine auxiliary proteins helping to assemble the metal cluster. How could DNA know in advance what to code? (p. 41)

He makes the excellent point that the Cambrian explosion saw the emergence of multiple phyla, or divisions of life-forms—more than we have now (p. 81). Why has the variety become stunted over time? If evolution leads to variety, then why isn't the world full of novel life-forms? Life, however, is an adaptation to environment. If the environment on this planet was capable of supporting a broader mix of life-forms in the past, then it's quite possible that our current world could lack a similar wonderful menagerie. The flux of Ice Ages and interglacials may have played a part in our modern epoch, for example, tempering a more diversified evolutionary procession. The Holocene, perhaps Anthropocene, is seeing that diversity cut back significantly: Humanity presents an environmental block on diversification as it domesticates nature to its own ends. Intelligent intervention, then, seems to work in the opposite direction to that being advocated in this book . . .

Another point that got me thinking was about the capacity for abstract thought among Neanderthals (p. 100). Does a lack of grave artifacts really indicate a lower level of development? We don't (generally) place items in coffins these days: Does that indicate that we are less-developed than our ancestors who did? Obviously not. There is an assumption that the development of religious thinking indicates abstract thought, and that, therefore, the absence of artifacts shows an absence of such development. But, perhaps Neanderthals just realized quite early on that there's no God? Perhaps theirs was a more sensible relationship with death. In terms of creativity, it is now recognized that Neanderthals painted art on cave walls 65,000 years ago, probably before humans did. Perhaps the Neanderthals taught the humans art. Who knows?

This book is chock-full of fascinating science. The author does not shy away from grappling with a high degree of complexity. Indeed, that is his very point. He leaves his hypothesis about the progenitors of this 'guided evolution' until the end of the book. This final section is speculative. It derives from Michael Behe's thesis about the fine-tuning of nature and intelligent design. Disappointingly, the nature of these designers is hinted at, but not stipulated. There is much about the mechanism of change; but by whom? This is where science needed to give way to philosophy. More searching questions needed exploring in this book.

The issue I have here is similar to skeptical arguments about God. Why do you need a middle 'man'? If the intelligent designers are carbon-based life-forms built of (roughly) the same biochemical constituents as us, then how did they independently 'evolve' to the level where they could do this themselves? Who intelligently designed *them*? If evolution occurred naturally for the intelligent aliens, then why not for us, too? Perhaps our designers are self-replicating robots with an artificial intelligence that has itself 'evolved' over millions, even billions of years? In which case one can only assume they were at least kick-started by a carbon-based life-form at some time in the past, and set free to continue their scientific experiments across the galaxy. Again, somewhere along the line, material beings had to evolve naturally first. Otherwise, we need God, or at least some kind of directing, intelligent *spiritual* force. In which case, we can just go straight to Creationism, and forget the science completely.

It strikes me that we really are 'just' manifestations of complex chemistry. If there's intelligent design, then it occurred at the blueprint stage of the universe: The rules, the laws, creating the opportunities for this complexity to emerge from carbon, hydrogen, nitrogen, and oxygen. Perhaps our universe is one of many—an infinite array within a multiverse; each with a different setup of laws and parameters. In which case, we live in *the* universe where such complexity is possible, and we have emerged on a planet where the conditions just happened to be right. Then, it's just down to statistics. No matter how remote a possibility, it occurs somewhere because there are so many potential planets, so many potential universes. That we are on the 'right' one is simply because this is the one where it happened, and our consciousness is available to record it.

I'm not ruling out alien intervention. There's a good chance that we have been visited in the past, and severely messed with. But that possibility can sit alongside the natural processes that increase complexity and functionality in response to environmental change. After all, somewhere, somewhen, it had to occur naturally to start with.

—**ANDY LLOYD**
Andy-lloyd@hotmail.com

# 11th EUROPEAN SSE CONFERENCE

*SOFIA CULTURAL CENTER*
*Helsinki, Finland*
*February 2019*

*For detailed information:*

*https://www.scientificexploration.org/*
*conferences*

# 38th SSE ANNUAL CONFERENCE, IN COLORADO

## Conference theme: Consilience

## June 5th to 8th, 2019

## OMNI INTERLOCKEN HOTEL, Broomfield, Colorado, USA

## conference rate $149/night

**For detailed information:  https://www.scientificexploration.org/conferences**

## Call for Papers in September

www.fundacaobial.com • fundacao@bial.com

FUNDAÇÃO

*Institution of public utility*

# *Funding for Scientific Research 2018/2019*

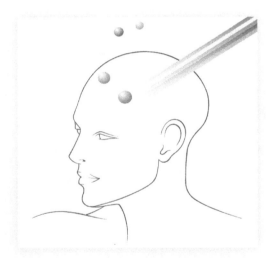

Through its Grants Programme for Scientific Research, the Bial Foundation is accepting applications of research projects in the areas of Psychophysiology and Parapsychology—projects from Clinical or Experimental Models of Human Disease and Therapy shall not be accepted.

Applications should be submitted in English by the 31st of August 2018, in accordance with the applic-able regulation and through the Bial Foundation Grants Management System (BF-GMS) available at

https://www.bial.com/en/bial_foundation.11/grants.18/grants_management_system.154/grants_management_system.a385.html

# Society for Scientific Exploration

## Executive Committee

Dr. Bill Bengston
SSE President
St. Joseph's College
Patchogue, New York

Professor Garret Moddel
SSE Vice-President
Dept. of Electrical & Computer Engineering
University of Colorado Boulder, Colorado

Dr. York Dobyns
SSE Treasurer
Lexington, Kentucky

Dr. Mark Urban-Lurain
SSE Secretary
CREATE for STEM Institute
620 Farm Lane, 115 Erickson Hall
Michigan State University
East Lansing, Michigan 48824
scientificexplorationsecretary@gmail.com

Dr. Chantal Toporow
SSE Education Officer
Northrup Grumman
Redondo Beach, California

## Council

Dr. Margaret Moga
Indiana University School of Medicine
Terre Haute, Indiana

Dr. Roger Nelson
Global Consciousness Project
Princeton, New Jersey

Marco Bischof
Berlin, Germany

John Valentino
Psyleron
San Francisco, California

Harald Walach
Europa-Universität Viadrina
Institut für transkulturelle Gesundheitswissenschaften
Postfach 1786
15207 Frankfurt (Oder), Germany

## Appointed Officers

Professor Anders Rydberg
SSE European Representative
Uppsala University, Uppsala, Sweden

Carl Medwedeff
SSE Associate Members' Representative
Livonia, Michigan

Professor Peter A. Sturrock
SSE President Emeritus and Founder
Stanford University, Stanford, California

Professor Charles Tolbert
SSE President Emeritus
University of Virginia, Charlottesville, Virginia

Dr. John Reed
SSE Treasurer Emeritus
World Institute for Scientific Exploration
Baltimore, Maryland

Professor Lawrence Fredrick
SSE Secretary Emeritus
University of Virginia, Charlottesville, Virginia

Professor Henry Bauer, *JSE* Editor Emeritus
Dean Emeritus, Virginia Tech
Blacksburg, Virginia

## BE PART OF AN ONLINE PK EXPERIMENT

We are recruiting participants for an online mind–matter interaction study of atmospheric turbulence. We encourage anyone, especially experienced meditators, to participate. A single session will take 15 minutes, and you may contribute as many sessions as you would like. Open to adults and United States residents only.

https://www.deltaaware.org/weather

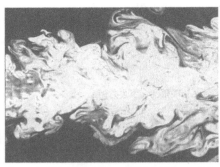

Photo credit: C. Fukushima and J. Westerweel, Technical University of Delft

Thank you. Dani Caputi, University of California, Davis, djcaputi@ucdavis.edu

# SSE ASPIRING EXPLORERS PROGRAM

itorious student research projects judged to be the most original and well-executed submissions in subject areas of interest to the SSE. A committee is in place to review all entries and determine the winners, who will receive awards of $500 each. One award winner will have the opportunity to present a talk describing the project at the SSE Annual Meeting, for which the Society will cover his/her registration fee. The other award winner will have the opportunity to present a talk describing their project at the SSE Euro Meeting, for which the Society will cover her/his registration fee. Submissions must be made per the guidelines and deadline as stated on the SSE website "Call for Papers" for the conference you are considering attending in order to be eligible for that year's prize for that conference.

If your paper is selected for the Aspiring Explorer Award, you will be either invited to present your talk at the meeting or able to submit your paper as a poster session. We are very excited about doing poster sessions now, so please let your fellow student colleagues and professors know about this.
https://www.scientificexploration.org/2019-conference

In addition, the SSE is also offering a 50% discount on future meeting registrations for any student member who brings one  student friend to our conferences (one discount per student). We are eager  to see student clubs or SSE discussion groups established at various academic institutions or in local communities. Contact us at sseaspiringexplorers@gmail.com to start your own group!

**C. M. Chantal Toporow, Ph.D.,  SSE Education Officer**

sseaspiringexplorers@gmail.com

# Index of Previous Articles in the *Journal of Scientific Exploration*

JOURNAL OF SCIENTIFIC EXPLORATION

## ORDER FORM FOR 2008 to current PRINT ISSUES

JSE Volumes 1–current are available FREE online at scientificexploration.org and
at http://journalofscientificexploration.org/index.php/jse/login
JSE Print Volumes 22–30 (2008–current) are available for $20/issue; form below.

| Year | Volume | Issue | Price × | Quantity | = Amount |
|------|--------|-------|---------|----------|----------|
|      |        |       | $20     |          |          |
|      |        |       |         |          |          |
|      |        |       |         |          |          |
|      |        |       |         |          |          |
|      |        |       |         |          |          |
|      |        |       |         |          |          |
|      |        |       |         |          |          |
|      |        |       |         |          |          |
|      |        |       |         |          |          |
|      |        |       |         |          |          |
|      |        |       |         |          |          |
|      |        |       |         |          |          |
|      |        |       |         |          |          |
|      |        |       |         |          |          |
|      |        |       |         |          |          |
| Postage is included in the fee. | | | | | |
|      |        |       |         | ENCLOSED |          |

Name

Address

Phone/Fax

**Send to: SSE**
**P. O. Box 8012**
**Princeton, NJ 08543-8012**

**Membership@
ScientificExploration.org
Phone (1) (609) 349-8059**

| ☐ | I am enclosing a check or money order |
| ☐ | Please charge my credit card as follows: ☐ VISA   ☐ MASTERCARD |

Card Number | Expiration

Signature

☐ Send me information on the Society for Scientific Exploration

## SOCIETY FOR
## SCIENTIFIC EXPLORATION

### GIFT JSE ISSUES AND GIFT SSE MEMBERSHIPS

**Single Issue:** A single copy of this issue can be purchased.
For back issues, see order form on previous page and Index.
**Price:** $20.00

**Subscription:** To send a gift subscription, fill out the form below.
**Price:** $85 (online)/$145(print & online) for an individual
$165/$225 (online only/print) for library/institution/business

Gift Recipient Name _____

Gift Recipient Address _____

_____

Email Address of Recipient for publications_____

☐ I wish to purchase a single copy of the current JSE issue.

☐ I wish to remain anonymous to the gift recipient.

☐ I wish to give a gift subscription to a library chosen by SSE.

☐ I wish to give myself a subscription.

☐ I wish to join SSE as an Associate Member for $85/yr or $145/yr (print), and receive this
quarterly journal (plus the EdgeScience magazine and The Explorer newsletter).

Your Name _____

Your Address _____

_____

**Send this form to:** *Journal of Scientific Exploration*
**Society for Scientific Exploration**
**P. O. Box 8012**
**Princeton, NJ 08543-8012**

**Membership@ScientificExploration.org**
**Phone (1) (609) 349-8059**

*For more information about the* Journal *and the Society, go to*
**https://www.scientificexploration.org**

# SOCIETY FOR
# SCIENTIFIC EXPLORATION

## JOIN THE SOCIETY AS A MEMBER

The Society for Scientific Exploration has four member types:

**Associate Member** ($85/year with online Journal; $145 includes print Journal): Anyone who supports the goals of the Society is welcome. No application material is required.

**Student Member** ($40/year includes online Journal; $100/year includes print Journal): Send proof of enrollment in an accredited educational institution.

**Full Member** ($125/year for online Journal; $185/year includes Print Journal): Full Members may vote in SSE elections, hold office in SSE, and present papers at annual conferences. Full Members are scientists or other scholars who have an established reputation in their field of study. Most Full Members have: a) doctoral degree or equivalent; b) appointment at a university, college, or other research institution; and c) a record of publication in the traditional scholarly literature. Application material required: 1) Your curriculum vitae; 2) Bibliography of your publications; 2) Supporting statement by a Full SSE member; or the name of Full SSE member we may contact for support of your application. Send application materials to SSE Secretary Mark Urban-Lurain, scientificexplorationsecretary@gmail.com

**Emeritus Member** ($85/year with online Journal; $145/year includes print Journal): Full Members who are now retired persons may apply for Emeritus Status. Please send birth year, retirement year, and institution/company retired from.

*All SSE members receive*: online quarterly *Journal of Scientific Exploration* (*JSE*), *EdgeScience* online magazine, *The Explorer* online newsletter, notices of conferences, access to SSE online services, searchable recent *Journal* articles from 2008–current on the JSE site by member password, and for all 25 years of previous articles on the SSE site. For additional new benefits, see World Institute for Scientific Exploration, instituteforscientificexploration.org

Your Name _____

_____

Email_____

Phone_____ Fax_____

Payment of _____ enclosed, *or*

    Charge My    VISA ☐    Mastercard ☐

    Card Number _____ Expiration_____

      **Send this form to:**
           **Society for Scientific Exploration**
           **P. O. Box 8012**
           **Princeton, NJ 08543-8012**

           **Membership@ScientificExploration.org**
           **Phone (1) (609) 349-8059**

*For more information about the* Journal *and the Society, go to*
**http://www.scientificexploration.org**

# JOURNAL OF SCIENTIFIC EXPLORATION
## A Publication of the Society for Scientific Exploration

## Instructions to Authors (Revised April 2017)

Please submit all manuscripts at http://journalofscientificexploration.org/index.php/jse/login (please note that "www" is NOT used in this address). This website provides directions for author registration and online submission of manuscripts. Full Author Instructions are posted on the Society for Scientific Exploration's website at http://www.scientificexploration.org/documents/instructions_for_authors.pdf for submission of items for publication in the *Journal of Scientific Exploration* (including "Writing the Empirical Journal Article." Before you submit a paper, please familiarize yourself with the *Journal* by reading JSE articles. Back issues can be read at http://www.scientificexploration.org/journal-library, scroll down to the Open Access issues. Electronic files of text, tables, and figures at resolution of a minimum of 300 dpi (TIF or PDF preferred) will be required for online submission. You will also need to attest to a statement online that the article has not been previously published and is not submitted elsewhere. *JSE* Managing Editor, EricksonEditorial@gmail.com.

**AIMS AND SCOPE:** The *Journal of Scientific Exploration* publishes material consistent with the Society's mission: to provide a professional forum for critical discussion of topics that are for various reasons ignored or studied inadequately within mainstream science, and to promote improved understanding of social and intellectual factors that limit the scope of scientific inquiry. Topics of interest cover a wide spectrum, ranging from apparent anomalies in well-established disciplines to rogue phenomena that seem to belong to no established discipline, as well as philosophical issues about the connections among disciplines. The *Journal* publishes research articles, review articles, essays, book reviews, and letters or commentaries pertaining to previously published material.

**REFEREEING:** Manuscripts will be sent to one or more referees at the discretion of the Editor-in-Chief. Reviewers are given the option of providing an anonymous report or a signed report.

In established disciplines, concordance with accepted disciplinary paradigms is the chief guide in evaluating material for scholarly publication. On many of the matters of interest to the Society for Scientific Exploration, however, consensus does not prevail. Therefore the *Journal of Scientific Exploration* necessarily publishes claimed observations and proffered explanations that will seem more speculative or less plausible than those appearing in some mainstream disciplinary journals. Nevertheless, those observations and explanations must conform to rigorous standards of observational techniques and logical argument.

If publication is deemed warranted but there remain points of disagreement between authors and referee(s), the reviewer(s) may be given the option of having their opinion(s) published along with the article, subject to the Editor-in-Chief's judgment as to length, wording, and the like. The publication of such critical reviews is intended to encourage debate and discussion of controversial issues, since such debate and discussion offer the only path toward eventual resolution and consensus.

**LETTERS TO THE EDITOR** intended for publication should be clearly identified as such. They should be directed strictly to the point at issue, as concisely as possible, and will be published, possibly in edited form, at the discretion of the Editor-in-Chief.

**PROOFS AND AUTHOR COPIES:** Authors will receipt copyedited, typeset page proofs for review. Print copies of the published Journal will be sent to all named authors.

CPSIA information can be obtained
at www.ICGtesting.com
Printed in the USA
BVHW04s2131141018
530163BV00013B/31/P